Reality outside the square
with
'Adremolin'
the last
'Wizard' of 'Oz'

Reality outside the square

with

'Adremolin'

the last

'Wizard' of 'Oz'

Adremolin

Copyright © 2024 by Adremolin.

Library of Congress Control Number: 2024911609
ISBN: Softcover   979-8-3694-9657-2
            eBook   979-8-3694-9656-5

All rights reserved. No part of this book may be reproduced or transmitted in any form or by any means, electronic or mechanical, including photocopying, recording, or by any information storage and retrieval system, without permission in writing from the copyright owner.

Any people depicted in stock imagery provided by Getty Images are models, and such images are being used for illustrative purposes only. Certain stock imagery © Getty Images.

Print information available on the last page.

Rev. date: 06/11/2024

To order additional copies of this book, contact:
Xlibris
AU TFN: 1 800 844 927 (Toll Free inside Australia)
AU Local: (02) 8310 8187 (+61 2 8310 8187 from outside Australia)
www.Xlibris.com.au
Orders@Xlibris.com.au
858032

Contents

Preamble vii

Foreword xiii

Part I: On My Favourite Subject Matter 1

Part II: Relatively-Real Thinking 33

Part III: The Atom-Core Model 56

Part IV: How Aware Are You? 67

Part V: Conclusion 87

Bibliography 103

Preamble

In the following story, I will tell how the universe is explained without the use of magic. This tale is in five parts. The first part is called "On My Favourite Subject Matter," and in this first part, I tell how I came to write my findings and give background information on standing, accepted earthly-theory. With it I prepare the reader for the second part: "Relatively Real Thinking." The second part of the story explains how matter and the universe are thought of by using "no matter" and "dark matter" in bond to create relatively real dimensional (spacetime) objects, yet the objects are not described with "infinitely-flat" dimensions but are described with relatively real curvature values. Starting up the matter production with the low-dimensional, object curvature–value neutrinos and ending the matter production with the high-dimensional curvature-value element, object number 118 of the periodic table of elements, in a ten and more solar mass sun. As well I will tell you how at the ninety-degree-angle mark, the dimensional (spacetime) duality separates "inversely directional," "without the use of 'math-e-magical' earthly trickery." I will show that singular "no matter" and singular "dark matter" are inverse to each other and so form matter only within a duality bond described as a relatively real dimensional (spacetime) object and modelled with a Pythagorean-like, right-angled *triangle*, which can be used to describe all relatively real matter. Also I will show that with this way of thinking a logical perpetual motion, one-way cycle can be formed, and with that perpetual motion one-way cycle, you can see that the universe never could have had a beginning and never will have an ending. The third part of this story is called "The Atom-Core Model," and this part of my story shows in my own way of thinking, which is still a little bit experimental—in line with our "Oz" way of living—how the leptons, (*lepto*: Greek for light because leptons are light particles) are able to fuse, causing the formation of the meson-baryon dualities (*meso*: Greek for medium; and *baris*: Greek for heavy) as quark symmetries creating the protons and the neutrons as two-up quark, one-down quark, and two-down, one-up quark units, which are then locked up in a bond within an electron cloud forming atom cores. The fourth part of this story is called "How Aware Are You"; this part gives an explanation on how the observable awareness in the universe can logically be created, based on the computational abilities that are only indirectly observable as quark symmetries yet match externally the relative observable electron-layered symmetries. I will show on my atom-core model that both the external and the internal symmetries are Fano Plane-like yet inversely reciprocal, and this inversely-reciprocal, inside-outside plus outside-inside computing ability that is observable on matter, for matter to potentially compute. I also show that all the DNA-based living objects in the universe may use this computer to a degree. The last part of my tale is called "Conclusion," where I sum it all up in relatively short form.

To
Believe
Means:

Not to know.

You can "believe" anything that you want to, without having to learn.

Yet for you to "know," you have to learn.

Foreword

I do have a memory of me as a little boy lying in the grass and facing the sky, looking at the clouds and beyond into the depth of the universe and wondering how all this "far out" and impossible-for-me-to-imagine universe containing billions of galaxies with billions of suns had started out of a big bang, a nil-point singularity. "How can that be?" And how I, that little nobody "Oz," boy wizard, think's about it and wonders: *How can it be?* Space is no matter and time also is no matter, yet old Albert Einstein said that matter is four-dimensional space-time. I don't know how many times I have tried to figure out this mystery as a little boy wizard, and I really tried hard with my mind, bending, twisting and mixing. No matter what I did, I never got anywhere with this flatland, space-time matter mystery. Yet my mate young Einstein showed me how to split Bier atoms into wasted space and wasted time. *How can that be?* I also remember as little boy contemplating going to find a cave high up somewhere in the Indian mountains and live there like the hermits in the fairy tales do and so will reach enlightenment and thereby understand reality. Yet instead of going "high" up in the mountains, I read mountains of books containing mostly unsatisfactory magical fairy tales until I got sick of reading confusing magical fairy tales and started to explore reality in many, many different ways, mostly by reading popular, and also not-so-popular, hard reading' science books and science-related material—and some interesting yet confusing books on spirituality that some gurus tied-in with findings of modern science in their attempt to make heads and tails of reality always hoping to find an answer that would satisfy my curiosity. Yet as interesting a time those books provided me, and as much fun my other explorations led me, not a single book, method, or person I had consulted along the way allowed me to understand all of reality in a meaningful way. None pointed even vaguely in the right direction, yet most of my mates pointed me to the pub, where the bragging young Einstein was spinning tales on split Bier reality. Yet I never gave up in my search, and my persistence paid off.

The dimensions that are used in theory to explain the universe are modelled confusingly in two ways: on the outside of matter, the dimensions your earthly wizards have modelled with ten singular space dimensions and one singular time dimension; on the inside of matter, you have modelled the dimensions as rolled up, forming the space-time bond that is now called the eleven-dimensional multiverse. Those dimensions are thought to be exactly as Euclid laid down his three space-dimensions about twenty-four hundred years ago as infinitely flat and in ninety-degree angles to each other. With it you have tried modelling the very, very small eleven-dimensional, space-time matter atom cores, as this was needed by the Theo-illogical mathematicians to allow them to cast the quark symmetry and the quark super symmetry spell, which allowed describing all atom cores. Yet this way of thinking to understand reality led to the wrong conclusions and therefore we have been searching for way too long in the wrong direction and so missed out in a logical way to understand reality. I will show with this tale how I managed to describe successfully reality in my "Oz-land" home with two none-dimensional, inversely directed singulars. One singular is called "no matter," and is like yin, and the other singular is called "dark matter," and is similar to yang. Both in bond are able to form equivalently-dimensional (spacetime) objects. This way of describing reality solves all of the confusing problems that wizards have encountered here on earth in the description of reality for at least the last twenty-four hundred years, and possibly even for the last five thousand. Yet with my way of thinking I found in Oz, I will show you in the following work how reality functions anywhere in the universe, in one singular story. In Oz, we use no magical trickery like wizards need in your earthly accepted standing theory to trigger the magical big bang, nor do we use in Oz the angel-like or the virtually-massed particle spell anymore, which Stephen Hawking here on earth had used in his attempt to evaporate matter via his virtual-particle spell. I vanquished all magical matter spells and replaced the old way of thinking with relatively real, relative dimensional (spacetime) objects. This allowed for a matter creation and matter separation process that does not need magical spells. In Oz we stick with simple Muggle logic and the laws of physics, and we in Oz stick to the here on earth, so elusive common sense.

I hope you will enjoy my Oz-Land tale.

—Adremolin, the last Wizard of Oz

To

be and not to be

is the beauty of

reality.

Terry Pratchett in *Raising Steam*, page 11, wrote that:

"It is hard to understand nothing, but the multiverse is full of it. Nothing travels everywhere, always ahead of something and in the great cloud of unknowing nothing yearns to become something, to break out, to move, to feel, to change, to dance and to experience—in short, to be something."

This way of thinking of reality is the result of infinitely-flatland dimensional thinking and is perfect for Terry's fantastic

DISCWORLD novels.

Yet observable reality is without this fantastic

Discworld dimensional worldview, and you will see that our reality is different.

Part I

On My Favourite Subject Matter

I have wanted to understand how reality works ever since the little boy wizard in me wanted to understand what he could not understand, which was how it was possible to make matter by using the three infinitely-flat space dimensions and the one also infinitely-flat time dimension without the magical attachments that I was taught at an Oz-land Catholic primary school. Therefore, I have been thoroughly studying, almost throughout my life, the new findings of science in books, magazines, and papers, hoping that these will reveal to me one day the making of matter via logical means. At retirement from work about twelve years ago, I was driving in my car along the pothole-riddled Oz-land Ocean Highway and had as usual my mind on matter and reality. We in Oz have a lot of spacious country, and I had a great, timely yet quite bumpy drive on that beautiful day that longed for a daydreaming, and so I let my mind wander and my thoughts run wild and free. And I contemplated that we still had not found the ultimate smallest particle, which we set out to find about fifty years before, and now fifty years on, I again wondered how it could be that matter is made with the ten infinitely-flat space dimensions and the one infinitely-flat time dimension, which we, since old

Einstein's general theory of relativity, have called a four-dimensional space-time. Nowadays we call Albert's relativity theory thought concept, an eleven-dimensional space-time multiverse. Standard accepted theories neither are able to give us a logical answer to that question nor do any of our theories enlighten us at all on how to separate these theoretical eleven space-time dimensional objects into ten singular space dimensions and one singular time dimension. Stephen Hawking, while denying the existence of God, had used alluring Theo-illogical math-e-magic in his evaporation model of a point-like space-time matter singularity, as he had used the illogical, angel-like, virtually-massed particle-duality spell in his theoretical attempt to evaporate the black hole's point-like matter singularity, yet neither did he separate matter into space and time nor did he use acceptable common sense logic to evaporate the mind fart of a black hole point-like matter singularity.

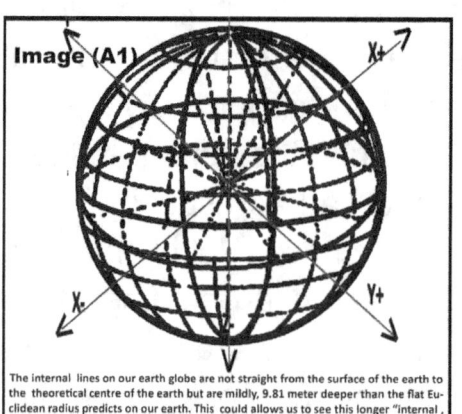

The internal lines on our earth globe are not straight from the surface of the earth to the theoretical centre of the earth but are mildly, 9.81 meter deeper than the flat Euclidean radius predicts on our earth. This could allows us to see this longer "internal , now "fat" time dimension as "Dark Matter-dimension" thereby we could understand "mass" yet this scenario still would fail in the description of eleven 'space-time' dimensions that we need to describe the super symmetries in all atoms.

On that day, along the bumpy highway road, I had a glimmer of a thought that the time dimension may possibly be associated with dark matter, and so I toyed on the road with the idea that if time, as a dark matter dimension, moves gradually into the three space dimensions. This should produce, with the continuous flow of the now dark time dimension into the three space dimensions the observable energy, also the observable mass and the observable gravity in matter. See image A1. The singular now dark or "fat" time dimension—I like this pun—could be seen as one internal dimension that gradually is getting deeper in internal curvature values, simulating this way gravity as it ages the object and the three flat space dimensions could be seen as the external relatively-flat dimensions forming the resulting inverse four-dimensional space-time. With it I thought we could possibly understand the flow of time, gravity, and the forming of matter yet not know how to separate matter other than with the extremely-confusing, virtually massed angel spell. I realized quickly that this inverse dimensional space-time duality mixture that I was trying to use was not much different to the one that is presently used, therefore it was likewise impossible to use for a model needed to describe reality with—now with one hypothetical fat time dimension and three hypothetical flat space dimensions. Particle mechanics had shown that it is possible mathematically, with eleven space-time dimensions, to model the atom's internal quark super-symmetries of the different atom cores; therefore I could clearly see that my resulting four-dimensional space-time model, now with one hypothetical fat time dimension and three hypothetically flat space dimensions could never form equivalent dualities with eleven space-time dimensions. So it was useless for me to hold on to my initially so exciting idea that the time dimension may be a dark matter dimension. Yet I gave my mind free reign on that drive, and it did not matter if my train of thought got me nowhere; I had fun thinking. Yet those thoughts never led me anywhere but home on that day, but they also never left my mind again. And gradually these thoughts guided me to understand the making of matter and led me to see how reality can be made without magic by following a strict logical thought train. We are now at the age of knowledge, and we have amassed within the last one hundred years a relatively truthful body of specialist-based scientific knowledge, which nowadays we all can access easily twenty-four hours each day, and I think that it is about time someone uses this vast amount of scientific knowledge, sorts out, and combines all the relevant related theories, carefully with logic and common sense, and assembles with all this knowledge a singular model of reality. This meant that I as well had this ability. Therefore I did, being retired, and interested in this field, what was needed, and by sorting through this vast amount of knowledge and twelve years of home-grown Muggle science in my office on the computer or in the studio painting or outside working in the workshop building atom-core model sculptures—and with the rigor of persistent truthful thinking (don't laugh) and also with the stubborn use of common sense (yes, I know common sense is to most by now almost mythical yet despite its myth I managed with it)—to produce a simple model that reflects the observable reality the way that reality is made and we observe reality to be made, not the way that we hope it to be made, with vivid imagination. I can clearly see that to obtain a logical meaning of reality anywhere in the universe, one must use the observable particle/antiparticle duality concept equivalently and apply equivalence to all that is relative. This equivalent duality principle can be observed in ancient artefacts all around the universe. For example in a forty-thousand-year-old caveman painting here on earth in the Kimberly mountains and a thirteen-thousand-year-old caveman painting in Argentina, there are depicted negative and positive handprints also in the yin-yang duality, which originated in old China, possibly back more than five thousand years ago to the particle/antiparticle dualities found in modern science. In Euclid's time, about two thousand four hundred years ago, you wizards thought that you lived on a flat disclike surface. This flatland worldview allowed Euclid to think of nothing more perfect then three infinitely-flat dimensions. With those three dimensions, Euclid modelled the universe and provided the elegance needed to lay down the foundation for the beautiful mathematics that you are still using today. Euclid's infinitely-flat dimensions bricked up the belief of a perfectly godly formed world, resulting in a universe where the earth, the sun, and the planets had been modelled as flat discs, using the perfection of the circle. Galileo Galilee showed you in 1609 that the moons, the earth, the sun, and the planets are not flat discs but are spheres. Albert Einstein in 1905 tried to explain reality with the three Euclidean, singular, flat space dimensions and his singular, flat time dimension. Einstein also tried to unite the three singular, flat Euclidean space dimensions with the one singular flat time dimension. With it you were meant to understand matter as a four-dimensional space-time. Edwin Hubble in 1924 found that the galaxies in the universe are receding from each other, and some of you wizards wrongly concluded later, with this observation, that the universe must have originated at one point. Arno Penzias and Robert Wilson in 1929 found background radiation in the universe wherever they pointed their dish antenna. In 1930 Chandrasekhar played with gravitational mathematics on his ocean journey from India to England and found that once the escape velocity of one object reaches the speed of light, it is impossible for a beam of light to escape this object, and so the idea of a black hole was founded. Fred Hoyle in 1949 wrongly assumed on the above observations and conclusions that the beginning of the universe happened about fifteen billion years ago. This event the theory says started with a hot big bang, a nil point singularity concept. The inverse to the hot big bang led to the theory of a cold big crunch in the far future, ending the universe. Penrose and Hawking stayed in line with theory, and in 1970 they showed that a gravitational collapse within the so-called black hole, containing a point-like space-time singularity is possible, using math-e-magic.

Yet I assure you that physically, point-like space-time singularities do not exist.

NASA observed that all vessels, like the Explorer vessels and the two Voyagers, are all slowing down on their outward journey. Voyager One for example, which NASA sent on September 5, 1977, with a relatively slow constant velocity travelling speed on an eighty-thousand-year-long journey to visit earth's next door neighbour sun three light years away, is slowing down as if some kind of friction is restricting its constant velocity. NASA said in 2012 on their website that this journey now may take twice as long as was initially predicted when it was launched, "If this vessel," they said, "ever reaches its destiny," and these days they don't directly mention the vessels slowing down, yet they do now say that the vessels had "encountered a new unknown force." The four infinitely-flat dimensions cannot be responsible for the slowing down of those vessels that NASA once was so sure would travel with the fixed constant velocity that the NASA geeks were able to achieve by slinging them from planet to planet, gaining in speed. The question arises: What is slowing these vessels down? Or what is that new unknown force? You observe here on earth that the super light leptons called neutrinos are able to pass mostly without interaction through your earth. Also you observe that the heavier objects, like the photons or solid objects like apples cannot do the same. Any observer of reality will agree that all matter objects in gas, liquid, or solid form cause travelling restrictions in the form of frictional interaction with other elements by exchanging energy, therefore causing wear and tear on the touching surfaces. On the other hand, the infinitely-flat dimensions of space and time do not, in your theories, cause resistance to the massive four-dimensional space-time; therefore, they do not interact at all, which means they cannot hit anything as they are not anything and so cannot exchange energy with matter, as both space and time in your earthly theories are free of matter. Since Einstein you wrongly accept that the space dimensions are subject to some change, and you have wrongly confirmed that the photons, which you also wrongly think travel in the flat space dimension, follow the thought of gravity-curved space values around suns and black holes, and by now you also wrongly accept that the same must be true for the flat time dimension. With your theories you can see that once these infinitely-flat dimensions are permanently joined as a four-dimensional space-time, those matter objects you observe then show the inverse to the above, where those four flat dimensions have no energy and have no gravitational curvature values. The joined four-dimensional space-time objects yet now all show energetic gravitational curvature values, which resulted in the spherical, astronomical, and atomic matter objects any wizard observes. These matter objects are by now modelled to have up to ten singular, flat, space dimensions in bond with one singular, flat, time dimension that is now called an eleven-dimensional space-time multiverse, and these all are still described as one massive matter object with Einstein's well-known equation $E=mc^2$. Yet no wizard is able to explain to us Muggle's with acceptable logic where the additional dimensions nor the energy comes from—neither the mass in the joining of the no energy infinitely-flat space dimensions and the no energy infinitely-flat time dimension, which is meant to cause the energetic eleven-dimensional gravitational curvature values in the observable, spherical, massive space-time objects. Not one honest Muggle understands with logic the arising multiverse, and neither do any of you wizards understand your confusing reality theories that you came up with after more than fifty years of costly research, other than with godly magic or similar fairy tales.

All these weird ideas and the so many confusing ways that we also once had used in Oz and you still here on earth boldly use today in the attempt to model reality made only my head spin, yet none ever enabled me to see the light. Yet this conglomeration of confusing theories together gradually made me see that the theoretical concept involving infinitely-flat dimensions causes not only confusion and chaos in the description of reality but also infinitely-flat dimensions cannot at all describe equivalently the dual curvature values that are observed in all matter. The (yin-yang) dualities reflect equivalently the (particle/antiparticle) dualities that are observed in particle research today. Those dualities form the bricks of reality and are everywhere and anywhere observed, yet those dualities are often perceived as singulars. You say, "This is good." Or, "This is bad." Or, "She is young and he is old." Or, "You are rich or you are poor." These qualities are never singular in the directly observable reality; therefore, those qualities are not absolute in any object in the observable universe. There is no object in the universe that is 100 per cent or absolutely beautiful, and there is no object in the universe that is 100 per cent or absolutely ugly. Likewise there are no absolutely good or absolutely bad objects or events anywhere in the universe. All consciously acquired and all relative accidently acquired experiences in all relatively aware objects in the universe are perceived, pending on the observer's point of view. The lion may think that today "I" would make a relatively good dinner; I on the other hand may think that the lion would make me a relatively good rug in the not-so-distant future. All things in the universe are dualistic and have two relative opposing gravitational curvature values. One side has a relative, positively-curved value forming the surface of the object and one side has a relative, negatively-curved value forming gradually with this value: the internal gravitational radius of any object. To prove this statement, I have come up with a thought experiment, and for this thought experiment, I will use the earth as what the earth is: a massive gravitational object. Let us visualize that we could cut off on the sphere of this earth one relatively-large flat surface, like a surface parallel to the tangent of the North Pole. This cut off surface would look, in principle, like a large, circular, flat disc, similar to what the hatched surface in image A2 looks like. The centre point of this disc is named point B and the external circumference of that theoretical flat disc is named A. This theoretically flat disc's surface is only relatively flat. This cut off surface on your earth will fill up with water and will form a relatively-deep lake or ocean, pending on size. The surface of this ocean would be described with a relative positively curved value; same as any other ocean or lake surface is described on this earth, and the bottom of this theoretically-flat ocean presents you with an inversely-relative, negatively curved value. You can see that the ocean would not only appear to be deeper at point B due to the observable positive surface curvature, it actually is deeper, as point B you also can see is closer to the centre of the earth, and therefore is deeper down than any point A is at the ocean's edge. On this theoretically-flat disc's surface, one has to walk uphill from the deeper centre point B to any higher point A at the ocean's edge. When you look at this thought experiment closely, it becomes obvious that there are no observable infinitely-flat lines, infinitely-flat surfaces, and no infinitely-flat rooms at all, in or on any object of the universe. There are only relatively curved rooms, relatively curved surfaces, and relatively curved lines observable on any object. All objects—neutrinos, electrons, atoms, rocks, moons, planets, suns, you and me—in the universe have mass and therefore are in possession of these inversely-positive, external and negative internal gravitational curvature values. See image A3.

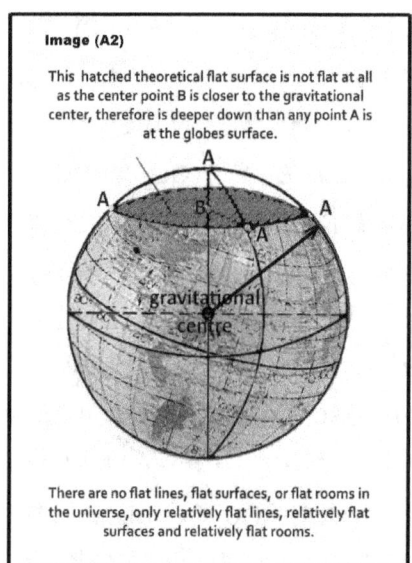

Image (A2)

This hatched theoretical flat surface is not flat at all as the center point B is closer to the gravitational center, therefore is deeper down than any point A is at the globes surface.

There are no flat lines, flat surfaces, or flat rooms in the universe, only relatively flat lines, relatively flat surfaces and relatively flat rooms.

The above thought experiment shows the negative curvature values that here on the earth are about 9.81 metres deeper down than the Euclidean radius predicts of the spherical earth. This thought experiment shows, and proves, that there are no perfectly-flat observable surfaces on any object anywhere in the universe. And it also proves that the absolute perfectly-flat, geometrical, and mathematical calculations based on the imagined infinitely-flat dimensions are not able to describe the by gravity curved positive-negative curvature values as you observe on earth to 100 per cent accuracy. That means that nowhere in the universe, but in theoretical mind farts, will you find a surface that enables you to use a compass and a straight-edge, the tools of the Euclidean trade, to draw like Euclid thought he did that would comply to 100 per cent with the Euclidean rules. Those flatland rules say that two times the radius is equal to the diameter of the drawn circle, yet on positive or negatively curved surfaces, you can see this is not so, and you also see that on the surface of the earth, you never observe the needed flat surface that Euclid had imagined to be able to apply the flat Euclidean rules using the Euclidean tools, as any flat surface on earth turns out to be a negatively curved valley. Therefore you also can see that neither is the equation true on any observable surface in the universe that says, "Diameter times pi is equal to the circumference," nor can you rely on the flatland equation that says, "Radius times radius times pi is equal to the surface area of that drawn circle." It is also obvious that you cannot draw with a 'straight-edge', parallel straight lines on any gravitationally curved object using the Euclidean

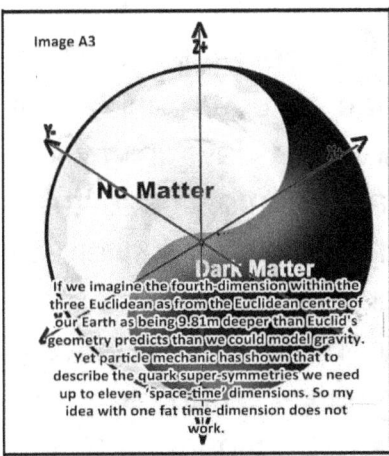

Image A3

If we imagine the fourth-dimension within the three Euclidean as from the Euclidean centre of our Earth as being 9.81m deeper than Euclid's geometry predicts than we could model gravity. Yet particle mechanic has shown that to describe the quark super-symmetries we need up to eleven 'space-time' dimensions. So my idea with one fat time-dimension does not work.

rules to 100 per cent accuracy. These rules and tools are only useable on the imagined, theoretical, infinitely-flat Euclidean ground and work very well on relatively-small objects, yet they cannot be used as they were thought of by Euclid—see image A4—on any surface anywhere in the universe, as all objects in the universe are objects with inversed dual gravitational curvature values. Twenty-four hundred years ago Euclid was convinced that the earth and the universe were flat, and whenever you are thinking like Euclid did, flat earth, as you do when you build or observe the relatively-small objects that Muggle's use in their daily lives, then the Euclidian geometry seems perfect, yet you can see that Euclid had invented those three flat space dimensions without any solid flat base with the erroneous idea that the earth and the universe is made on flat surfaces, yet you have just seen with my above thought experiment that those three flat space dimensions do not exist in observable reality. I had to use, back in my working days in my Oz-land home, these three theoretically infinitely-flat space dimensions in many, many relatively-small technical and constructional drawings

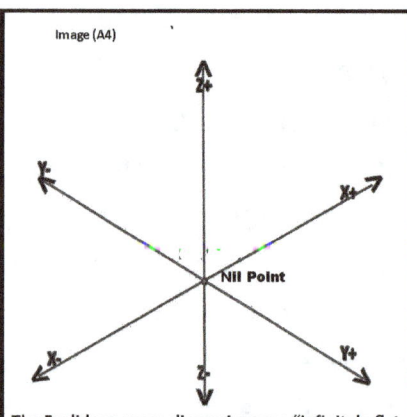

The Euclidean space dimensions are "infinitely-flat. The Nil-point is seen in standard theory as: the "Big Bang" event. At this event: x+ and x- plus y+ and y- plus z+ and z- are leaving with the speed of light in all outward directions and time is parallel to the three Euclidean space dimensions.

on many relatively-small sheets of paper. Never was I able to draw perfectly-flat objects in the office nor was I able to build perfectly-flat objects on the building sites nor in the workshops while using the Euclidean space dimensions. The Euclidean space dimensions are quite useful when you are building small objects and you calculate those small objects with flatland mathematical thinking as those small objects have, relatively speaking so little internal gravitational curvature values so as to be almost infinitely flat. And none bothers with their tiny internal gravitational curvature values when building small objects. The surface of the earth is way larger than your small toys are, and the earth is obviously not even close to being called an infinitely-flat object as you all can feel the negative gravitational curvature value of the earth, which attracts your bodies towards its internal gravitational centre with nine point eight-one metres per second per second at ocean level. For any building that you are constructing here on earth, you have to use the earth's positive surface curvature and call it a flat horizontal floor level. You have to also use the vertical plumb line for the upright walls. You

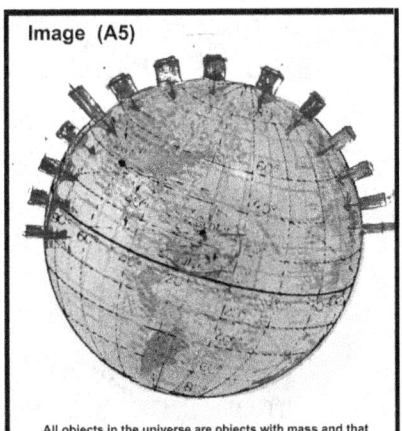

Image (A5)

All objects in the universe are objects with mass and that means they have gravity and as we have seen gravity causes dual curvature values. On our earth we use the earths surface curvature as floor level and the vertical lines as up-right walls. It is easy to see that Euclidian geometry is unusable on a large scale. Squares are not square, parallel lines are not always parallel and if you imagine using a compass to draw from the north pole a circle that describes the equator than you see quite clearly that the value of pi is not valid.

call them both squares, as both lines are at right angles to each other, yet you can see that vertical plumb lines are not parallel. See image A5. This shows to anyone who is able to visualize that the earth is a gravitational sphere and not an infinitely-flat object in the Euclidean sense and that on any gravitational object, there are only relatively curved surfaces lines and rooms, never perfectly flat lines or flat surfaces and neither are there perfectly-flat three-dimensional rooms. The same is also true for the vertical plumb line; this line too is not a perfectly-straight line down to the centre of the earth but is always a relatively curved line, pending on mass. This line is always a little bit longer than the Euclidean geometry would predict. Your mathematics is based on the three perfectly-flat Euclidean space dimensions; therefore, you can clearly understand now using my above thought experiment that the three imagined, Euclidean infinitely-flat dimensions are a theoretical figment of the mind. Yet you

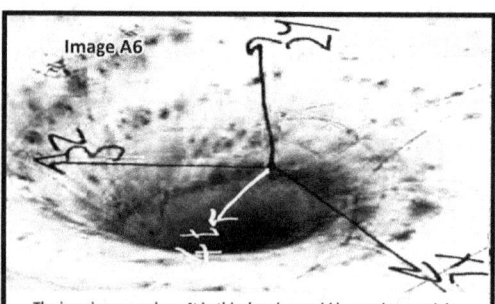

The imaginary numbers 4t in this drawing could be used to model a four-dimensional geometry where the fourth dimension, the time dimension could act as one internal dimension and with it we may be able, to degree, to model gravity and the resulting curvature values that are observable in all matter objects. Yet it takes a lot of imagination to translate this highly curved internal dimensional geometry to the observable 'space-time'.

also can see that these three dimensions are so excellent to use in flatland mathematics and in the production of the so many relatively-small objects you use in your daily lives so that one forgets to keep in mind that this theoretical perfection is only arising out of Euclid's three-dimensional, infinitely-flat way of thinking, not at all out of a modern logical, observing, thinking mind. With the need to model, mathematically in particle mechanics the quark super symmetries with more than four space-time dimensions that Einstein had thrown together as matter objects, you enlarged this theoretical four-dimensional model again in the same way old Einstein had enlarged Euclid's dimensions with the time dimension. Some mathematician had enlarged the Euclidean dimensions with the imaginary number dimension. See image A6. You are neither with the imaginary number dimension nor with the imaginary time dimension able to draw or build objects, yet math-e-magicians are able to use the imaginary number dimension in calculations. The same also applies to the theoretical ten space dimensions and the one theoretical time dimension that bond to form the eleven space-time dimensions needed to describe the super symmetries in the different atom cores. For a while even I in Oz thought this concept might work, and it does to a degree model the relatively-observable quark reality as it suggested that it is possible within degrees that one could imagine the added dimensions as facing the same way as the imaginary number dimension, from the nil point in ninety-degree angles inward and downward. Let the imagined dimensions gradually roll up deep down on the inside of the atom cores, and as Calabi-Yau provided, using the concept of string theory, a model to simulate internally the rolled-up dimensions. With this way of thinking, one could imagine a universe using theoretical mathematics, which allows rolling up, internally, the needed dimensions and so show how the observable matter symmetries could be formed. Yet with this confusingly-theoretical math-e-magical spell, one cannot explain physically this matter-making process nor can one manage to explain with logic how one got those extra dimensions—nor can one explain how eleven infinitely-flat dimensions are able to roll up only on the inside causing the massive curvature value, quark super symmetries within the atom cores, yet are thought to be infinitely flat on the outside of matter, causing the mind-blowing multiverse. Neither can one with logic explain how point-like matter singularities form. Also you can see that the more space dimensions you imagine, the less equivalently the resulting space-time objects become. This all is very confusing. Yet you all can see that you need dualities to explain the observable universe. No matter how hard wizards are trying to use the imagined flatland dimensional dualities of many infinitely-flat space dimensions and one infinitely flat time-dimension to model this universe into a logical, simple-to-understand concept, the infinitely-flat dimensions Euclid had invented—and Einstein also despite his relative thinking and you wizards as well after more than one hundred years of intense research still stubbornly hang on to—make this duality journey not only very confusing, it makes it impossible. Einstein tried to describe the observable reality with four space-time dimensions, and you know that he had failed with those four dimensions. Using eleven space-time dimensions where the dimensions are internally tightly rolled up in super symmetry and are externally infinitely flat within a multiverse does also not explain with logic how mass and energy are formed. Since Albert's general theory of relativity, wizards have been trying with four infinitely-flat dimensions to model the universe in two unlike ways. In one version the three infinitely-flat space dimensions and the one infinitely-flat time dimension, you wizards have modelled as independent and as separate dimensions, and those separate dimensions leave, in theory, the nil-point singularity at the hot big bang. At first in the inflationary face of the big bang theory, the three infinitely-flat space dimensions and the one infinitely-flat time dimension inflate, parallel to each other, the universe faster than the speed of light, being no matter, and from the expansionary face of the big bang, the three infinitely-flat space dimensions and the one infinitely-flat time dimension are then described as cooler and as being fused to form the now four-dimensional space-time. Those now massive four-dimensional space-time objects are thought to expand with the speed of light in all outward directions and are thought to model this imagined reality now from the past hot nil-point state, which is described as the beginning of the universe, to the future cold nil-point state, which is meant to model the end of the universe. Yet you all can see that this way of thinking causes quite easily confusion and not just to anyone, as you know what old Einstein thought: reality is relative. Yet old Einstein also thought that space is inherently flat and old Einstein also thought that God does not play with dice; this itself shows relative confusion. They say the apple never falls far from the tree. No wonder young Einstein back home in Oz was also confused and so tried to split Bier atoms. Therefore I think the reason why old Albert and his first wife, Mileva Maritsch, did not apply the otherwise used equivalence principle also to dimensions when they were modelling the general theory of relativity as they possibly could not weigh down their conscience with relative dimensions. Albert and his first wife, Mileva, chose to stick with the infinitely-flat dimensional concept of Euclid's thinking and applied the same concept to the invented time dimension; therefore Albert and Mileva had to form the four-dimensional space-time with a ratio of three space dimensions to one time dimension in bond. We see that those (three-to-one) four-dimensional space-time objects can never be equivalent at all. So you can clearly see by now why both minds were unable to describe relativity in one singular theory. Albert later, in 1949, came up with a second theory to describe what could not be described with the general theory of relativity; the second one is called the specialized theory of relativity. Albert tried till the end of his life in

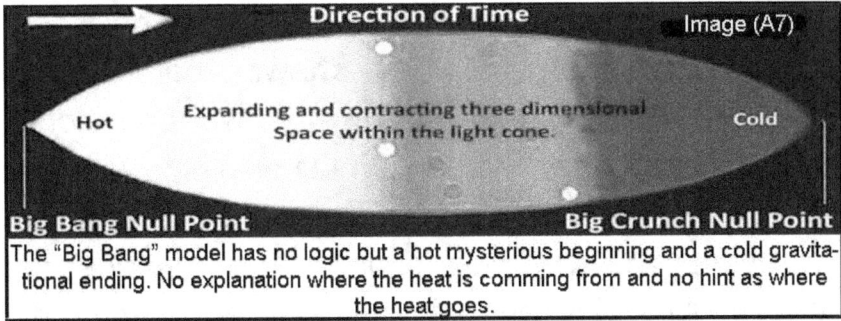

The "Big Bang" model has no logic but a hot mysterious beginning and a cold gravitational ending. No explanation where the heat is comming from and no hint as where the heat goes.

1951 to unify both theories but failed to do so. The massive four-dimensional space-time objects that we have seen with my thought experiment above all show you equivalent positive and negative gravitational object curvature values. Those dualistic gravitational object curvature values cannot be separated, not even in theory, into the three singular space dimensions and the one singular time dimension in the accepted model of the universe. Hence the confusingly invented hot big bang nil-point singularity, the illogically cold big crunch nil-point singularity—see image A7—and the attempted angel-like, half-massed virtual particle exchange, which Stephen Hawking had used in his evaporation model of the Hawking-Penrose point-like space-time singularity.

You imagine and draw and build your buildings, weapons, and toys, only in theory with the use of the three space dimensions. When I asked myself back in old Oz, "How is it possible to do drawings in two infinitely-flat space dimensions and build in three dimensions while using four-dimensional materials? I also got myself an answer. The perfectly-flat Euclidean geometry clearly showed that no three-dimensional object could occupy the two- or the one-dimensional space. The same must also hold with four flat dimensions. Therefore no four-dimensional object can occupy the one, the two, or the three-dimensional space. My quick and simple answer to my own questions was that all drawings and all buildings are made with and on the surface of the relatively curved four-dimensional space-time objects and never in infinitely-flat land—only the absolutist-based theories and the arising mathematical imaginations are. On the other hand, wizards also know that one cannot mathematically model the observable matter with Einstein's four-dimensional space-time, as nowadays particle mechanics and astronomy have shown that to mathematically model the strong and complex internal gravitational curvature values associated with the quark super symmetries in the atom cores or to model the massive internal gravitational curvature values associated with suns and black holes more than four dimensions are needed. To do that Kaluza-Klein suggested at first five, then six, then eight space-time dimensions. You then enlarged again his idea to now eleven space-time dimensions, and Calabi-Yau imagined those dimensions rolled up, simulating, internally, the high curvature value super symmetries of the atom cores, using string theory, yet you kept the same many dimensions on the outside of the atoms infinitely flat, resulting in the theory of a multiverse. Wizards know very well that this confusingly imagined theoretical duality of rolled-up versus infinitely-flat dimensions helps modelling the inside of the atom cores using the alluring magic of theoretical mathematics and therefore justifies math-e-magically their use in a theoretical solution to describe the inside of matter. Yet you wizards have no such magical luck with solid matter nor with the arising infinitely-flat eleven dimensions, which are meant to form the outside of the atom cores, other than with fantastically smelling mind-farts; those are now being used in the media for many, many confusing yet magical fairy tales. These different viewpoints do use a duality of sort, yet the duality of rolled-up versus flat dimensions are not able to model energy or gravity. To do that you need, in bond, inversely-equivalent dualities. As you can see, no matter how tightly and confusingly you may bend many imaginary infinitely-flat dimensions, the resulting messy space-time reality is meant to form multidimensional point-like singularities and also are thought to be made of mind-blowing, angel-like, virtually massed matter dualities, where infinitely many virgin angels have to sit on the tip of some horny devil's dick. This is never logically understandable nor is it inversely-equivalent and therefore not useable to model energy nor gravity. What you have produced in the last twenty-four hundred years with the absolute and infinitely-flat dimensional thought concept to describe your reality are smelly Theo-illogical mind farts, and those farts you have been greedily absorbing, digesting, and re-farting again and again in many different holy yet quite smelly and very bloody notes around the world.

The dualities that wizards have found in particle-mechanic research form the base of reality. In your daily lives, it is easy to understand that observable reality has these equivalent massive gravitational curvature values, and pending on the size of these massive gravitational curvature value objects you are able to work on, you can see and have to consider within degrees these gravitational curvature values in planning and construction of relatively-small or relatively-large projects. For example: by stringing up a clothesline here on the positively curved surface of earth, you can observe and prove on the downward bow of the clothesline that a negative gravitational curvature value is present. When shooting a gun on this positively curved surface, you have to adjust pending on the bullet's speed, for the negative gravitational curvature value that attracts the bullet with 9.81 metres per second per second towards the centre of your earth. When you are launching a shuttle into orbit or when travelling to the moon, you need to consider the massive negatively curved gravitational attraction of the earth to lift off, and you have to consider the less massive negatively curved gravitational attraction that is on the moon—only about 1.63 metres per second per second—to land. On the way back home, the inverse must be considered . When you are building a huge concrete dam, then you have to consider the positive gravitational curvature value of the earth's surface curvature to calculate the larger amount of steel and concrete that is needed on the earth's surface than you would need for the same job on infinitely-flat ground. You also can easily see that the rules of the imagined Euclidean space geometry apply only to Euclid's imagined infinitely-flat reality and change once they are applied to the dual gravitational curvature values that you observe on any object in the universe. You can clearly observe on the earth's surface curvature that Euclid's geometry behaves differently than it does in his imagined infinitely-flat land theory. With infinitely-flat land theory, you know and can see that the equally angled, or isosceles triangle, with three sixty-degree angles when enlarged, changes only into a larger equally angled isosceles triangle, yet it still forms a triangle with three sixty-degree angles. On the

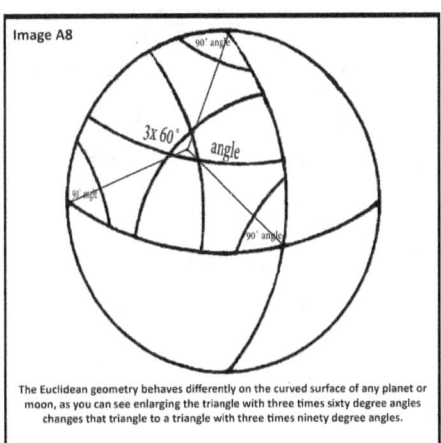

The Euclidean geometry behaves differently on the curved surface of any planet or moon, as you can see enlarging the triangle with three times sixty degree angles changes that triangle to a triangle with three times ninety degree angles.

other hand, you also observe that on the surface of this earth that this in small scale now gradually changes into a positively curved isosceles triangle with larger angles when you do the same operation—see image A8—like three ninety-degree angles forming a quarter of a sphere and larger angles up to the point where the three sides of that enlarged isosceles triangle each form one third of the equator and so form a quite complicated isosceles triangle, with three-thirds of the equator's circle as the three sides of this triangle. This enlarged triangle you can see is now covering the surface of one half of the earth's sphere, with three possible sections. This enlarged, complicated isosceles triangle gradually changes again with further enlargement, forming a complete sphere with a negatively curved isosceles triangle at the opposite side of that sphere as a gradually closing opening in three possible inversed sections. The value of pi is not the same when measured on gravitationally curved objects as it is in theoretical flatland, and you can clearly see that vertical parallel lines do not exist at all on any surface anywhere in the universe. Horizontal parallel lines are forming circles. You have also seen with my thought experiment that the internal surfaces of any gravitational object are always valleys and never flat surfaces, so you can see that the Euclidean space dimensions are a theoretical mind figment and do not exist at all anywhere in the universe; therefore, they can only be used anywhere in the universe in relatively small scale or in huge mind farts. The theoretical Euclidean dimensions provided the magical foundation for Euclid's geometry and his geometry also provided the base for the imaginary objects that you magicians are able to conjure with fancy flatland math-e-magical spells, thereby ignoring common sense. Never were those infinitely-flat dimensions the arising geometry nor is the resulting flatland math-e-magic able to describe matter with 100 per cent accuracy.

It should be crystal clear to you wizards by now that all objects in the universe have equivalent dual gravitational curvature values, and it should also be equally clear to you that with infinitely-flat dimensional thinking, you will never find satisfactory answers to explain the observable dual gravitational curvature values equivalently nor are you able to explain the energy dualities—neither mass nor the observable perpetual one-way motion of birth and death in all the relative, observable objects. You wizards have accepted since old Einstein that reality is relative; therefore your building blocks and your thinking of reality must also be relative. Relative shows dual qualities. You know that relative is not quite this and you know that relative is not quite that. The infinitely-flat dimensional way of thinking is like mathematics: precisely flat and fixed, causing fixed results and flatland infinities; therefore both cannot be used to describe the relative observable dual gravitational curvature values of objects with 100 per cent accuracy, as 100 per cent accuracy does not exist anywhere in the universe. Your theoretical description of reality is using these imagined, invisible, infinitely-flat dimensional tools to describe reality, yet you all observe that all matter in the universe is only relatively observable. So how can you, within acceptable logic, describe and model relative observable—and in theory multidimensional—objects when, with acceptable logic, you only can use three space dimensions, and these you only can use in relative small scale. You have to completely ignore the large flat-dimensional part of this fancy multiverse, except in theory or in media fairy tales. The infinitely-flat time dimension in bond with the three infinitely-flat, Euclidean space dimensions already needed magical imagination to pretend that one is actually able to visualize the creation of the four-dimensional space-time as an energetic massive object. And to think of the relative, confusing duality of the eleven infinitely-flat dimensions, which wizards use in particle mechanics, as infinitely flat on the outside of matter yet use internally as tightly rolled-up dimensions to enable the mathematical description of all elemental cores and think of the confusingly resulting flat, ninety-degree-angled dimensional external multiverse with a logical thinking mind is mind-boggling. I am sure it is obvious to you by now that with the theoretical concept of infinitely-flat space dimensions, one cannot think, draw, or build objects in more than three space dimensions. This we are only able to do to the relatively-small objects that we Muggle's handle quite well in three dimensions. Wizards have tried for a long time to describe, with their relatively small-scale observations, the very, very large gravitationally curved objects in the universe, and they also have tried with those same relative small-scale observations to describe the very, very tiny gravitationally curved particles that you wizards observed in the accelerators' target chambers and failed to do it with elegance and logic. Your magical theories are based on the very, very strong belief that the observable objects in the universe are formed out of the smallest matter building block, and you thought that this ultimate smallest matter building block, which some actually here on earth call the God particle, can be found by smashing larger particles into smaller ones, thereby making all matter particles godly. Those technological feats were very exciting yet also produced confusion as you wizards persistently observed that the objects in the universe do not yield the singular, smallest, godly particle that you were looking for, yet you observed only small particle/antiparticle dualities, just like yin-yang. Yet you wizards still stubbornly hang on to the absolute and infinitely-flat way of seeing reality. To be able to describe all the relative observable objects in the universe, Muggle's have to have relative tools to do the job. We have known since Heisenberg that measurements are relative, and Muggle's persistently observe that dimensions are never infinitely flat nor are they infinitely large. Absolute dimensional objects do not exist but in mind farts.

What Really Are Dimensions?

When I was a little boy wizard in my Oz-land primary school, I believed that the dimensions were as I had been told: invisible and infinitely flat. Those to me then were understandable as they were explained as Euclidean space dimensions, being x, y, and z. This was sort of logic to me as I was able to draw and calculate with those dimensions the objects that I dealt with. The same dimensional length of x, y, and z describe in geometry and in mathematics the cube. Yet the teacher at school also told us that the three dimensional Euclidean universe is a spherical object, that all outward reaching directions in the Euclidean space are infinite and so must have the same length and therefore the three dimensional Euclidean space describes the universe as an infinitely-large sphere, not a cube. When I tried to argue with him on that point, I was told not to cause trouble in class and shut up. The teacher then said that since Albert Einstein, we also have to include time as a fourth dimension, and with those four dimensions together, we can describe all space-time objects. The fourth dimension, the time dimension, was never understandable to me, as I could never draw or build with this new dimension.

This bothered me, so you may understand why this time dimension was a constant confusion and pain on jobs that I had worked on. The time dimension never showed up in the drawings nor when needed on the job. Yet now when I look in the mirror, I am almost horrified to see what the elusive virtual time dimension has done to me.

So yes, I asked myself: "What really are dimensions?"

And my answer to my own question was: dimensions are the measurable properties of objects!

Dimensions are not invisible, imaginary, infinitely-flat things. You all observe that the objects in the universe are only relatively measurable. Think again of Werner Heisenberg and the uncertainty principle. The measurable dimensions of all objects in the universe are also only uncertain, *never* certain. That means that to understand reality without the use of magic, you have to change from the invisible, infinitely-flat magical worldview that you wizard's use and have religiously pounded into most of us with confusing yet beautifully imagined magical mind farts to the now relative, observable, relatively real dimensional worldview that we common Oz-land Muggle's use and with common sense created relatively real reality. With this little change in thinking, I will show you that it's not the invisible four or more dimensional, infinitely-flat—nor the rolled-up eleven—dimensions nor the elusive point-like particles that you have searched in vain for within the last one-hundred years but the visible, relatively real, low-to-high dimensional, space-time objects are the building blocks of reality. Trying to describe the relatively observable objects in this day and age with the use of imaginary point-like singularities and imaginary infinitely-flat dimensions is *magisterium staff*, as you all can clearly see that with those confusingly-imaginary, infinitely-flat dimensions or point-like singularities, you are not within acceptable logic, meaning common sense. They defy observation, slap common sense, and only cause confusion, conundrums, and chaos in the description of reality. Your standard accepted theories and the observable reality are at odds with each other as one cannot do the mathematically predicted flat-out job on any dual-inversed, equivalently curved surface. This is exactly what we are having as playing ground in the universe. The playing ground is not wrong the Theo-illogical, flatland dimensional way of thinking that wizards persistently hang on to when trying to describe the relative is.

If you now hang on to my circular yet strict outside-the-square, Oz-land thought train, then I promise you will get the perpetual spin of many lifetimes.

Part II

Relatively-Real Thinking

My careful and logical conclusions that I drew back in Oz to get out of this infinitely-flat dimensional mess was to go back to basics and scrap the big bang theory, scrap Euclid's infinitely-flat land thought concept, and scrap Einstein's thought of the four or more dimensional space-time and start to describe reality anew; this time with the use of logic and not any Theo-illogical mind farts. My observation was that the planet Oz shows undeniable equivalent dual gravitational curvature values. Remember my thought experiment and the observation that the space-time matter duality can never be separated into the three or ten singular space dimension and the one singular time dimension. The observation that the galaxies must have about 90 per cent dark matter to hold the suns in galactic orbit made me think one day of a way to possibly describe reality with equivalent dualities. By thinking of dualities, I thought that if dark matter could have its inverse, say no matter, and I attribute the no matter value to the space dimensions and the dark matter value to the time dimension, then we could possibly describe with the gradual merger of no matter for three or more space dimensions with dark matter for one time dimension: the four or more dimensional space-time objects that we need to describe the atom cores. With this train of thinking, as so many times before on my journey, I was wrong. 'No matter' cannot be three or more space dimensions and 'dark matter' cannot be one time dimension as those dimensions are the cause of the space-time confusion. If we think this clearly, then we will understand. Dark matter is singular, and as a singular, it is not tangible, which means singulars are only indirectly observable. We have not seen any dark matter directly, and we have not seen any no matter, yet we know that dark matter must be present in the galaxies. Scientists have used the concept of no matter to expand the universe faster than the speed of light in the inflationary face of the big bang theory. If I apply the commonly used flatland dimensional way of thinking, as wizards still do, to my hypothetical no matter, then we have my hypothetical no matter, and we have the indirectly observable dark matter and still have absolutely nothing happening as they will never join in this way to form objects with any magic-free logic. To make something happen with logic, we need things in motion, and to move things we have to explain and join no matter with dark matter in such a logical way that allows the gradual making of matter with both singulars. You will see that with this way of imagining the universe, a perpetual motion one-way cycle is logically forming, and I will show in this perpetually one-way cyclic motion that no matter is in the upward, outward motion and inversely that dark matter is in downward, inward motion. Both being singular, they are not dimensional, so they are only indirectly observable. Yet both bond because the dimensional (spacetime) values needed to build and describe the observable objects in the universe--and as you now see with elegant equivalence. To write this relatively-real duality bond, I have placed the written dimensional (spacetime) object within parenthesis. I did this so not to confuse you, or anyone, with the commonly used four or more dimensional space-time. The commonly used four or more dimensional space-time cannot be separated into the three or ten singular space dimensions and the one singular time dimension, hence those hypothetical dimensions ended up as a hypothetical eleven-dimensional, point-like matter mind fart. With that point-like matter mind fart formed the Theo-illogical, multidimensional space-time singularities that evaporate religiously the point-like matter mind farts with angel-like, virtually-massed particles, which now must be travelling quite faster than the speed of light, and we find at that speed, outside the black hole, a new angel-like, virtually-massed partner particle to form a new duality for the evaporation spell of the black hole's point-like singularity to work.

I placed in my matter model the dimensional curvature value of the matter-object duality in front of the parenthesis, resulting in a relatively real low-to-high dimensional (spacetime) object. In my story I have renamed the black holes to avoid confusion. I call them dark suns as my dark suns do not contain point-like space-time singularities; my dark suns are relatively real objects that form and exist between relatively real (low to high) dimensional (spacetime) object curvature values. I have ascribed the status of my hypothetical no matter to yin to be able to form the inversely directional status of the indirectly observable dark matter, which I used for yang. I used the (yin-yang) image to allow the visualization of the no matter / dark matter bond as one relatively real observable object. See image A9. The singular none-dimensional no matter and the singular none dimensional dark matter are not directly observable and are without form, as singulars not able to make (yes/no) decisions and therefore cannot think. Computing, thinking, and relatively real objects need the energy dualities that can be described with the bond of kilometres and seconds within a now relatively real, Pythagorean-like right-angled triangle, yet this triangle is now an underlined <u>triangle</u> and is describing

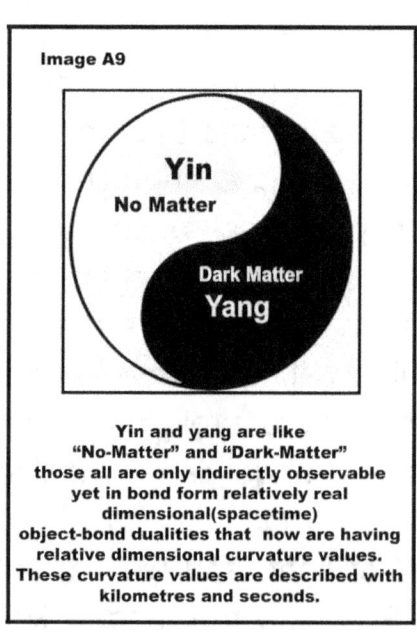

Image A9

Yin and yang are like "No-Matter" and "Dark-Matter" those all are only indirectly observable yet in bond form relatively real dimensional(spacetime) object-bond dualities that now are having relative dimensional curvature values. These curvature values are described with kilometres and seconds.

low-to-high dimensional (spacetime) object bonds. My relatively real dimensional (spacetime) object-bond <u>triangle</u> too cannot be separated into many singular space dimensions and one singular time dimension, as my Oz-land findings show that there are no singular space dimensions nor are there singular time dimensions at all to even consider this mind fart. Yet we can see that my relative dimensional (spacetime) object-bond <u>triangle,</u> which is made of the no matter / dark matter bond, does separate the relative dimensional (spacetime) object in the relative, indirectly-observable parts of the universe deep down within a dark sun's container, where at the ninety-degree angle mark, the energetic dimensional (spacetime) hypotenuses' duality gradually folds up its dimensional (spacetime) trigonometry, forming inversely directed singulars of no matter going up, or outward, and dark matter going down, or inward.

Relatively real in my work means that any object in the universe is just like we observe; they all are gradually conceived and then gradually born when relatively young in one form or another. They then all gradually change in dimensional (spacetime) object curvature values in one form or another. Just look in the mirror and you will see what I mean: we all grow taller, fatter, older, greyer, and frailer. We all gradually die relatively old in one form or another. You can clearly see that you and all objects in the universe are only relatively real, never forever, yet most of you insist that the objects—and you—in the universe *all* are real. Yet you and the objects in the universe are not absolutely real, as absolutely real would mean that the objects in the universe do not change and therefore do not age, so would exist forever. You observe however that you and all things around you in the universe do age and do change; also you observe that you and all things around you in the universe do not exist forever. Only the nothing is unchanging and forever, yet the nothing do not exist at all. You have never observed an object that will exist forever, and you never will. Relatively real reality is only ever relatively observable and that means relatively real. Relatively real objects are with gradually changing uncertain boundaries. Relatively real observable objects in the universe always have relatively-young, relatively-low-dimensional, gravitational object curvature value beginnings. This also means that all relatively-observable objects in the universe always do have relatively-old, relatively-high-dimensional gravitational object curvature value endings, just the same as you observe all the objects in the universe, including suns, planets, moons, and you and me. No object in the universe is absolute, infinite, unchanging, or never-ending. The singular kilometres of space and the singular seconds of time do not exist. The things that exist for a while and make up reality are the relatively real observable dimensional (spacetime) objects that result out of the no matter / dark matter duality bond, causing the relatively real low-to-high dimensional (spacetime) objects that all are born and all will die. This forms the perpetual one-way cycle of the universe. So you see the universe has many, many relatively-low-dimensional object curvature value beginnings and many, many relatively-high-dimensional object curvature value endings without the Theo-illogical, so-very-hot big bang and the so-very-cold big crunches. We see also that the universe had no absolute beginning and has no absolute ending. Yet as I have to start that cycle somewhere to begin this short tale with, I now start the cycle in the same way that all the relatively real objects in the universe start: as a conceived (o) relatively real, low-dimensional (spacetime) object that gradually changes and matures to become the baby object, which at birth has relatively-mild object curvature values; in this case about 0.01 to 0.50 dimensional (spacetime) object curvature values as mass when this object exits at the event horizon of the dark sun's container as a neutrino baby lepton object and inflates with mildly faster than the speed of light. So I think, as we know that the lepton neutrinos are quite lighter than the photon lepton bosons are, the theoretical no matter unit value up to the ergo sphere's diameter now acts as the energetic outward expansion speed and is then checked in by its inversed inwardly directed dark matter unit value speed. Both of those bonded, inversed, theoretical values describe the relatively real volume, the relatively real mass, and the relatively real gravity value of the object and also give insight to the travelling speed of the neutrino, which now inflates the universe from the inside of reality to the outside of reality in all outward directions mildly faster than the speed of light and creates the universal background. This neutrino lepton gradually ages as it travels outwardly, acquiring gradually incoming dark matter and causing gradually more mass within the neutrino, thereby it gradually changes the dimensional (spacetime) trigonometry of the neutrino and with it the curvature value of the neutrino. The neutrino gradually changes externally into a smaller outside diameter yet forms internally gradually a deeper gravitational radius and so changes the initially almost-flat neutrino to a higher-dimensional energy lepton neutrino object and gradually changes to the quite massive lepton object: the electron. Those then gradually change via the same ageing process to super-high-energy electron clouds, containing within the core of this cloud individually higher-energy electrons that, as they individually shrink in external diameter, gradually crowd closer to each other. This way they form denser electron-cloud regions, causing within the core of these clouds higher dimensional curvature value cauldrons. Therefore electrons begin to fuse in those internal, mildly-higher curved cauldrons to the low-dimensional (spacetime) hydrogen atom. With the perpetual and gradual downward motion of more incoming dark matter is produced. All young matter objects age and age causes, as we see, mass and gravity going through stages. At first this process is only gradually forming low-dimensional hydrogen, which produces relatively-large, first-generation proto suns, which in their lifecycle are causing newborn galaxies yet gradually ending this journey in the relatively large—from ten to billions of solar-mass value, dark sun container cauldrons, which now deep down are gradually reaching at the ninety-degree angle mark the impossible ten-dimensional (spacetime) object curvature value, as matter at this state separates inversely to no matter and dark matter. No matter is going upward and outward and dark matter is going downward and inward, causing again this inversed motion duality the collision of both singulars with the upward, outward motion of no matter and the downward, inward motion of dark matter, also causing deep down conception, which again grows to be one (low) dimensional (spacetime) neutrino, which is leaving the event horizon and inflating the universe in all outward directions, causing again objects between relatively-low and relatively-high dimensional (spacetime) object curvature values and gradually is causing again no matter going up and dark matter going down.

In my Oz findings, the no matter units are invisible in nought-degree angles, and the dark matter units are invisible in ninety-degree angles. Both these unit values in duality form a relatively real dimensional (spacetime) hypotenuse describing with it object dualities, now with relative, indirectly-observable no matter values and relative, indirectly-observable dark matter values that in bond describe now all the relative, directly-observable, relatively-flat to relatively-fat dimensional (spacetime) object dualities in the universe. I have started the perpetual spin of many, many lifetimes with the relatively-low dimensional curvature value objects called neutrinos. Neutrinos are the light members of the lepton family. *Lepto* is Greek for light, and I am going to explain to you how leptons can be formed and described with a now underlined, yet in principle same as a Pythagorean triangle in detail below. The neutrinos exit the event horizon and inflate their potential no matter value, up to the ergo sphere's diameter. See Image A10, where this almost-flat, inflating expansion speed is held in check with the inversely directed dark matter value speed—both in bond describe the neutrino as a relatively real object. Deep down within the relatively real dark sun container, the newly conceived neutrinos theoretical no matter value is still heavily confined within the narrow macaroni-like container, and so when the newborn neutrino is exiting the relatively-narrow event horizon, it is thereby inflating almost flat out up to its potential its no matter value, where the opposing, inwardly-directed dark matter value speed checks this rapid, outwardly inflating expansion speed at the ergo sphere's outside diameter and so the newborn dimensional (spacetime) duality neutrino leaves possible mildly faster than the speed of light, as the neutrino is definitely a lot lighter than a photon is. So I think the neutrino, being flatter than the photons, can travel a little bit faster than the photons can. The neutrino needs not to be an angel-like, virtually-massed particle to leave the dark sun container. In this way the neutrinos are inflating the universe's background in all outward directions, from the inside of reality to the outside of reality with almost flat-out, low-dimensional neutrino curvature values. This creates the relatively-low-dimensional object curvature values, the universal operating room, thereby gradually creating the background radiation that makes up and sustains the universe in a perpetual, one-directional cyclic motion. This neutrino diameter arises out of the dark sun's internal gravitational radius times two. The (no matter / dark matter) object is describing in bond

The Ergo Sphere describes the external diameter of the Dark Sun container object and also describes the external diameter of the new born baby neutrino object. The internally curved radius is the result of gravity. This curvature makes relatively real objects from relatively 'flat' matter objects to relatively 'fat' matter objects.

the (space value and the time value) of the newborn neutrino as one object. And now you can see that the young neutrino is inverse to the old dark sun container. The internal gravitational radius of a ten solar mass dark sun that has a hypotenuse with a curvature value of about 89.9° degrees has a gravitational radius of about two-hundred and ninety-nine thousand nine-hundred and seventy-two kilometres deeper down than the Euclidean geometry would predict of a newly born, ten solar mass sun. See Chandrasekhar forming with this relatively (high) dimensional (spacetime) object the funnelled breeding ground of new relatively real (low) dimensional (spacetime) neutrino, now with inversed dimensional gravitational object curvature values to the dark sun's dimensional gravitational object curvature values. My relatively real neutrinos objects to the old fat-dimensional dark sun container object curvature values I have modelled using the principal of the abstractly formed Pythagorean triangle, yet now I use this Pythagorean triangle as an underlined relatively real curved triangle. Pythagoras had described his abstract triangle with one side named (a) and one side named (b). Both of those sides are at ninety-degree angles to each other. You know that both sides form the resulting abstract hypotenuse side, named (c). In this right-angled triangle, (a) squared plus (b) squared is equal to (c) squared, and the root of (c) squared is (c). Now I use this triangle as a relatively real triangle that has one relative, indirectly observable side named no matter and one indirectly observable side named dark matter. Both values are not directly observable yet facing always in ninety-degree angles to each other. This causes the gravitational hypotenuse and gradually changes the now relative, directly observable, dual dimensional (spacetime) curvature angle. This describes, on the yellow, one-way brick road of reality, a relatively real observable yet gradually changing, ageing, and relatively real dimensional (spacetime) object. So again you see no matter^2 plus dark matter^2 equals matter^2, and if you now take the root of the squared dimensional matter, it becomes the relatively real curved gravitational hypotenuse, describing with it the relatively real dimensional (spacetime) object—now as a bond of kilometres and seconds, the substance and the energy needed to be relatively real matter, using only simple Muggle logic. This start this cycle with a neutrino that has relatively real (low) dimensional (spacetime) gravitational object curvature and ends the matter cycle at the moment that the relatively (high) dimensional (spacetime) object's curvature value has

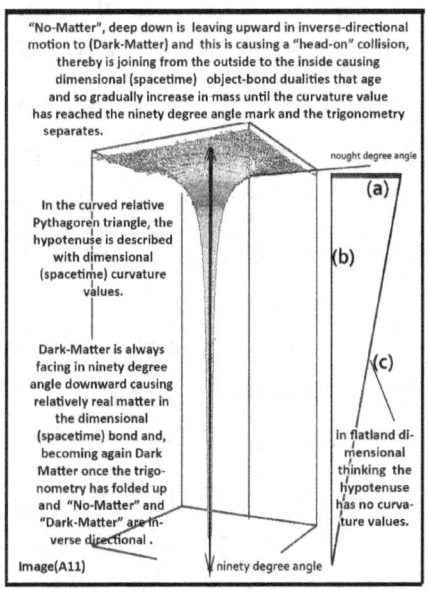

reached internally the curvature value needed to collapse this triangle's relatively real status. That happens once the hypotenuse within this container reaches the ninety-degree angle mark, and at that ninety-degree mark, the hypotenuse now folds up this relative triangular object into inversely directed singulars. Deep down within this container, as you can observe in the diagram—see image A11—the gravitational curved value causes the hypotenuse to only gradually reach the ninety-degree angle mark, not suddenly collapses to point-like singularities the bond that describes the relatively real hypotenuse into inversely directed singulars of no matter going up and dark matter going down. The dark matter motion is always only from the outside to the inside, observable on any relatively real object on the spherical surface of the object as being in ninety-degree angles and dropping toward the inside core, describing with the dark matter units, now in confinement, relatively real observable matter and with it the massive negative dimensional object's curvature values and its positively-curved external values. This describes the gravitational attraction value here on earth at about 9.81 m/s². The no matter unit value in nature is inversely to the dark matter unit value in nature. Therefore the no matter unit is doing the inverse of the dark matter unit. That means that the no matter unit is moving upward or outward, inversely to the dark matter unit, which is moving downward or inward. The no matter unit is in upward-outward motion, or from the inside to the outside, causing, still deep down at that relative ninety-degree angle mark, the head-on collision with the still incoming downward motion of the dark matter units and forming a confined relative-motion duality as a dimensional (spacetime) object now with a dark matter unit potential as mass and a no matter unit potential as explosive outward energy, therefore $E=mc^2$.

The newborn neutrino is leaving the dark suns ergo sphere mildly faster than the speed of light as the neutrino is lighter than the photon. The neutrino travels 299,982 km/s², which I think is needed for the neutrino to escape the dark sun's gravitational attraction, which is set at the speed of light with 299,972 km/s². The underlined triangle is describing both objects. The neutrino is described with this triangle as a relatively-flat (low) dimensional (spacetime) object, and the dark sun is described with this triangle as a relatively-fat (high) dimensional (spacetime) object. You can see that this triangle is describing the birth of the (low) dimensional (spacetime) objects, and you can see that this triangle is also describing the ending of the relatively-fat (high) dimensional (spacetime) objects that I called dark suns. Yet you can also see with this relatively-curved underlined triangle that only the hypotenuse of that triangle is relatively real observable as it is an object with relative, observable, dimensional (spacetime) object curvature values, thereby the hypotenuse is not suddenly collapsing into a point-like space-time singularity, as one would logically think if the hypotenuse were an abstract flat line in abstract flat space, yet the relatively real hypotenuse has dual curvature values that internally are negatively curved, and so we see the internal hypotenuse's value does not suddenly collapse to a point-like singularity. The triangle is still retaining on the gradually shrinking, external, positively-curved ergo sphere containing the event horizon's surface, opening a diminishing value of the squared no matter value and a diminishing squared dark matter unit value; therefore this triangle also retains a gradually diminishing internal hypotenuse with a lessening negatively-curved value. You can see that the end product of my relatively real (high) dimensional (spacetime) object story are cleanly yet gradually separating at the only theoretically (10) dimensional state into inversely directed singulars doing the up and down or the in and out thing. In this way again, they become new almost-flat (low) dimensional (spacetime) objects. Never do dark suns have point-like

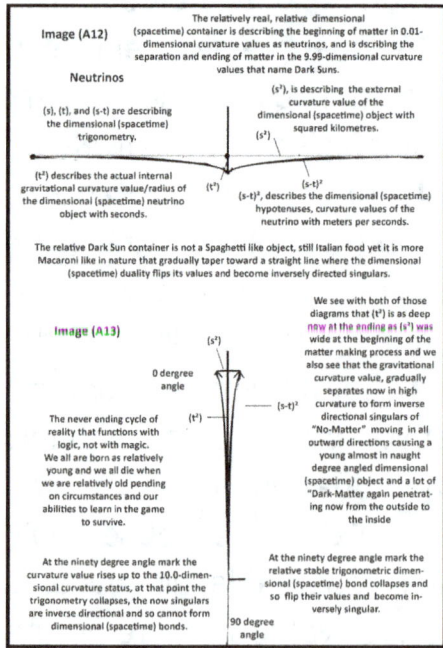

space-time matter singularities. The neutrinos are born by exiting the event horizon and inflating to relatively the same outside diameter as the ergo sphere's outside diameter of the old, relatively real dark sun, yet the neutrinos have a very, very flat, now almost in nought-degree-angled gravitational object curvature value. On the other hand, the old dark suns a very, very fat, almost in ninety-degree angled internal gravitational object curvature, starting at the event horizon and going down. See image A12 and A13. The neutrinos have a relatively real (low) dimensional (spacetime) object curvature value and so the neutrino has a very shallow dark matter value or gravitational internal radius. The relatively-fat (high) dimensional (spacetime) objects have a relatively-deep dark matter value, pending on mass, describing the internal gravitational radius, which is for a ten solar mass sun about two-hundred ninety-nine thousand nine-hundred and seventy-two kilometres deeper down than the Euclidean geometry would predict of a newly born, ten solar mass sun. Describing with this dual curvature value the massive energy potential of the relatively-fat (high) dimensional (spacetime) object and describes with this value also the possibly mildly faster than the speed of light's travelling value the neutrinos need to escape this object. Both objects, the relatively-flat (low) dimensional (spacetime) neutrinos and the relatively-fat (high) dimensional (spacetime) dark suns are very alike, yet the neutrino object is very, very low in gravitational curvature values and the relatively-fat (high) dimensional (spacetime) dark sun is very, very high in gravitational object curvature values; one has a very shallow internal gravitational cauldron and one has, inversely, a relatively-deep internal gravitational cauldron. One is very large in outside diameter and one is very narrow in outside diameter. With the forming of that relatively real underlined triangle, you now can explain the production of the matter object dualities equivalence. With this triangle you can describe gravity, understand energy, and comprehend the forming of mass with logic. With this triangle you can observe that dark matter is always going down, always from the outside to the inside, and with it you also can observe the perpetual change within the young, relatively real flat (low) dimensional (spacetime) objects to the old, relatively real high (fat) dimensional (spacetime) object curvature values. And also you can logically describe with this triangle the gradual separation process of the (fat) dimensional (spacetime) matter object curvature values deep down at the ninety-degree angle mark, where you see the triangle is gradually separating the bond to inversely directed singulars of no matter going up and dark matter going down, producing again with this inversely directed motion relatively-low (flat) dimensional (spacetime) object curvature values and a lot of singular dark matter going down. The newly conceived (flat) dimensional (spacetime) object is now potentially the newest, lightest, smallest matter particle in the universe as all other objects can be modelled from this. I explain this in the next chapter.

To summarize: The singular dark matter is going down, always from the outside to the inside and no matter is always travelling from the inside to the outside, potentially flat out always and upward-outward. This causes, still deep down in the dimensional (spacetime) container, only from the outside to the inside and only at this point, the possible merger of no matter with dark matter via the head-on collision. This produces the newly conceived (low) dimensional (spacetime) baby neutrino, now being an object with relatively flat dimensional (spacetime) object curvature values. You can see it is now a relatively real dimensional object that is created with equivalence and with dual gravitational curvature values, now having a little matter. It is therefore described with almost flat dimensional gravitational object curvature values. See image A14. The relatively-real, yet almost macaroni-like, relatively-fat funnelled (high) dimensional (spacetime) container object allows the now conceived (low) dimensional

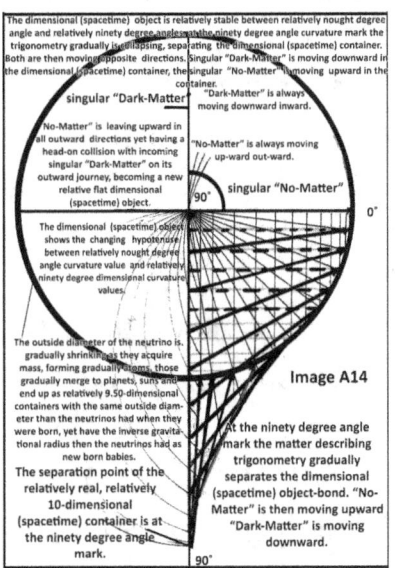

Image A14

(spacetime) neutrino to grow gradually on its upward-outward journey, thereby it ages, causing within this container the maturing of the neutrino up to a 0.50 dimensional (spacetime) object curvature value. The neutrinos are exiting at birth in all outward directions the dark sun's event horizon with these highly squashed curved values, and so they are inflating up to their theoretical no matter unit value, which is reached at the ergo sphere's object diameter, possibly mildly faster than the speed of light. They are then checked in in this outward speed with the opposing, inversely directional, theoretical dark matter unit value. With both values in bond the neutrinos are now described as relatively real outside the diameter, the neutrinos have relatively real mass, the neutrinos have relatively real energy, and also the neutrinos are travelling at a relatively real speed, as one relatively real object, inflating now the universe's background with energetic, relatively real, flat dimensional object curvature values; therefore, they are pushing the galaxies with relatively-flat dimensional objects from the inside of reality to the outside of reality, causing the galaxies to recede from each other. The neutrinos then gradually get older with incoming dark matter and so gradually acquire more mass and age, thereby shrinking individually and gradually in external diameter yet increasing individually and only gradually in their internal curvature values. This causes gradually higher energy lepton states called high-energy neutrinos. Those then gradually change to lepton electron states, which again gradually are enriching the universal background with higher-energy electron levels. Those higher-energy electron levels continue to age and gradually change to even higher-dimensional electron curvature values and so become gradually smaller in external diameter yet heavier and deeper in internal gravitational radius. Therefore, you see, as the volume of each individual high-energy electron becomes smaller in external diameter, their internal values become deeper and their energy increases. They all are crowding closer together and so are forming a denser electron cloud that gradually gets highly enriched with high-energy electrons, causing a huge cloud-like strange region in the universe that is gradually increasing in mass yet is gradually shrinking in external diameter up to the point where internally, in the higher curvature value-forming cauldron, the electrons are so saturated with energy and so desperately in need to give off photons to lighten their heavy loads—yet they are unable to do so due to being confined within the gradually deepening cauldron. Therefore the electrons are heating up to the point where they gradually do become less repulsive to each other. This enables the electrons gradually to touch each other—both being very hot they are less repulsive to each other—and so when touching they give off their internal surplus energy by emitting, or dumping, photons into the created negative spaces that had formed when the electrons had tried to repel from each other. They now fuse with the inversed curvature values of the photons within the negative spaces to the positively curved, negatively charged electron skin, causing with it the quarks in quasi confinement to become the meson-baryon dualities, with it forming units of protons and neutrons that are now shielded with the fused electrons in confined cloud form, forming all atoms in the universe. And now you also can understand that the relatively-fat (high) dimensional (spacetime) container object separates inversely at the same rate from the inside of the (high) dimensional (spacetime) dark sun to the outside, as the ninety-degree angle penetrating downward-inward flow of dark matter is, from the outside of the relatively real low to high dimensional (spacetime) container object, to the inside. That means that the separation process of this relatively-fat (high) dimensional (spacetime) container object takes inversely as long as it had taken to form these relatively-fat (high) dimensional (spacetime) container objects. This shows that Einstein's equivalence principle is retained in all that we observe within the universe, dimensions included. This also tells us that the observable universe is not and cannot have infinite space values nor can it have infinite time values but is always a relatively real, relatively measurable, and continuously changing yet finite dimensional (spacetime) universe containing a multitude of spherical yin-yang objects with dual gravitational curvature values between relatively-flat" and relatively fat gravitational object curvature values.

These objects are now perpetually powered by the never-ending yet only one-way motion of invisible singular mindless no matter and invisible singular mindless dark matter, creating in bond the never-ending yet only relative dimensional, relative mind-full, and relative mind-less objects that exist between the relatively-flat dimensional object curvature values in almost nought-degree angles as neutrinos and the relatively-fat dimensional object curvature values in almost ninety-degree angles as dark suns. Both the null-dimensional and the ten-dimensional curvature values do not exist at all. We see that at nought-degree angles, we have no hypotenuse and at the ninety-degree angle mark we see the hypotenuse describing the dual dimensional curvature values folded and is no more, and so the no matter unit is freely going up from the inside to the outside, causing deep down the head-on collision with the still incoming dark matter units going down. This causes a relatively real dimensional (spacetime) object duality with now dual-dimensional gravitational curvature values in bond as massive objects that have the ability to age and so increase gradually in mass. This mass gradually causes again the separation of this duality. And so again no matter is going up and dark matter is going down, causing the little bang the two inversely directional singulars are having, and this little bang caused the energetic conception of a new, relatively real dimensional (spacetime) object. No complicated magic needed.

Part III

The Atom-Core Model

This image below (A15) shows the simple (two up, one down quark plus two down, one up quark) symmetries as a (meson-baryon) duality that make up the proton and the neutron. This describes and forms the resulting electron-cloud symmetries of my deuterium atom-core model. This is still a little bit in the experimental state yet is in principle not at all that different to the one that wizards are using in standard theory and have investigated on the inside and on the outside for at least the last one hundred years. See image A16. As you will see, I have only changed the way that wizards interpret the quark status. So far in standing theory, wizards describe the quarks as "elusive particles," yet in my atom-core model, the quarks are not at all elusive but are quasi particles as you can never observe quarks directly. Treating quarks as quasi particles allows flipping the currently modelled six elusive particles that make the proton-neutron

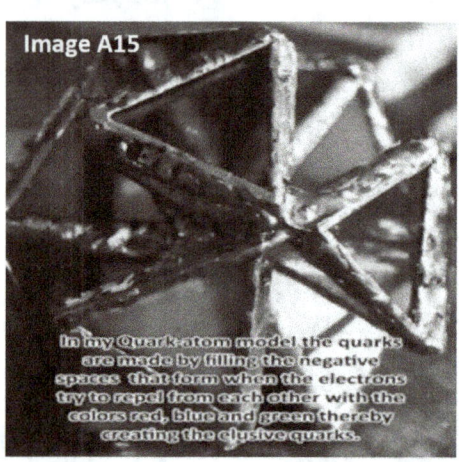

bond and use instead the six elusive particles and four relative, observable electrons as we can fill the resulting six negative spaces that form when four electrons try to repel from each other with dark matter, causing fusion. This forms the six quasi particles as two up, one down plus two down, one up quark and with it explain the logical production of the quarks without confusion. I have introduced my idea of electron fusion to build atoms, and with my atom-core model, I can provide a quite elegant backup. My quark atom-core model explains and shows how the electron's gravitational curvature values gradually increase with age to higher-energy electron levels thereby gradually forming huge high-energy electron clouds. Those huge high-energy electron clouds gradually gravitate and so form internally higher dimensional curvature values than the outside dimensional gravitational object curvature values and so form internally a dimensional cauldron where once the electrons are forced via gravity to touch each other. They become very, very hot so have less repulsive force to repel each other on

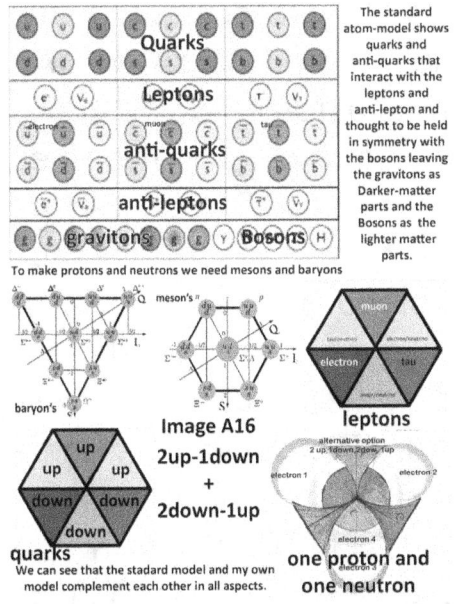
Image A16

the cauldrons inside. They then fuse in symmetry, forming the three quarks needed to build the one-proton element hydrogen, creating with it the first-generation suns that I call relatively-large proto-suns. Those first-generation proto-suns do what all suns do: they produce atoms, pending on their cauldron's internal dimensional curvature values. The first-generation suns, or proto-suns, are super huge in external diameter yet only have a relatively-shallow low-dimensional internal gravitational curvature value cauldron formed of quite hot yet relatively low-dimensional electrons. They therefore only produce relatively-low-dimensional, one-proton hydrogen atoms. At the end of the first-generation proto-sun's life cycle, the first-generation proto-suns then go supernovae, flinging the produced hydrogen in all outward directions possible and leaving behind a relatively small amount of deuterium and a relatively large electron cloud, which forms relatively large galaxies or even may form cluster galaxies that now also each contain up to 97 per cent of dark matter. The produced hydrogen atoms in the new galaxies then do the same as the electrons in the background did and age via incoming dark matter, getting gradually older and more massive. In this way they gravitate and form relatively-large hydrogen clouds, producing via gravitation and age second-generation suns with a higher dimensional internal gravitational curvature value. Those then fuse the hydrogen elements in the larger/deeper cauldrons to the first proton-neutron atomic element called deuterium with now two up, one down plus two down, one up quarks and produce within the gradually increasing cauldron's curvature value all elements that follow, up to the element iron. Suns with higher dimensional cauldrons are needed to make higher dimensional atoms—two, three, etc. ... up to relative ten-solar-mass suns. Deuterium with one proton and one neutron in my atom model forms the base core element, where all other elemental quark super symmetries can be modelled. See image A17. My atom-core model is gradually growing similar in appearance and shape to the Calabi-Yau model, yet my model does not need to roll up infinitely-flat dimensions to cause quark symmetries. My atom-core model becomes gradually a multitude of interconnected dimensional (space-time) quark symmetry and super symmetries that twist and roll up, thereby forming the higher-dimensional curvature values on the inside and the lower-dimensional curvature values on the outside. They hide the higher dimensional inner symmetries, therefore the observation of the fatter, higher inner

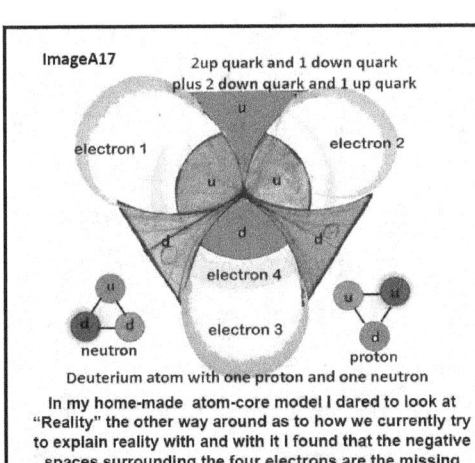

ImageA17   2up quark and 1 down quark plus 2 down quark and 1 up quark
electron 1   electron 2
electron 4
electron 3
neutron   proton
Deuterium atom with one proton and one neutron
In my home-made atom-core model I dared to look at "Reality" the other way around as to how we currently try to explain reality with and with it I found that the negative spaces surrounding the four electrons are the missing parts that allow a coherent logical description of reality.

dimensional symmetries is difficult from the flatter outer points of view. Super symmetries are quite similar to mirrored isosceles triangles, which face relatively angled to each other, therefore observing this model causes easy confusion. So far I have constructed the very basic few atom cores and its isotope cores, and I also have attempted to build a larger atom-core model and found that it becomes complicated and confusing very, very fast, not to build the model but to mentally follow the resulting complex super symmetries. Glass or Perspex I have found so far is a good material to observe those gradually rolling up symmetries, up to a point, yet I see that in principle this atom-core-model building method may not only be capable in theory to describe the forming of the neutrino leptons as baby matter, the model is also capable of explaining how it is possible to fuse the resulting leptons in higher-dimensional curvature values to higher-dimensional elements, thereby creating the beautifully interconnected quark symmetries and quark super symmetries that are stable up to the point where the resulting object's internal gravitational curvature value symmetries are only producing relatively unstable isotope symmetries with more and more neutrons. My atom-core model can also explain why in these different fusing processes, from element to element, a gradually lessening, relatively-large amount of energy is released.

In theory I can fuse electrons to relatively real, flat (1 to 2) dimensional (spacetime) atoms. I don't know if it is possible to test this first-generation super-high energy electron fusion here on earth. Yet a practical model describing the quark symmetry is possible, and with paper plus paint or stained glass plus metal or Perspex plus silicon or glue, anyone can model this event. See image A18.

Image nine (A18)
In my atom model, the quarks are quasi particles, those are made by filling the negative spaces that form, when four high energy electrons try to repel from each other by filling the negative spaces with the colours red, blue and green.

By using glue and four paper hexagrams as pretend high-energy leptons, I am able to show the units that form the two up, one down quark plus two down, one up quark symmetry. Those quark symmetries form the meson-baryon duality, which causes one proton and one neutron, and those basic symmetries form the basic shape of the idealistic atom core. I modelled the quarks in my atom-core model by ascribing to the resulting negative spaces that form, at the gluing (fusing) process of the paper or glass electrons, the colours red, blue, and green, and so I am able to show the elusive quarks in a similar way we observe in the particle accelerator's target chambers the only indirectly observable elusive quarks. Once you smash the glue that holds together the symmetry that make up the quarks, thereby forming the protons and neutrons, you only observe lepton parts scattering; you never see the quarks. See image A17. The same holds in my atom-core model. Once I smash, so to speak, the glue that holds together the paper or the glass electrons, the paper electrons then separate into different lepton parts and the negative spaces disappear and that makes the quarks as elusive as dark matter. The perfect atom has one proton and one neutron, and this shape should be the base symmetry to describe all the other elements. The deuterium element has that simple (particle/antiparticle) one proton, one neutron atom-core symmetry and shows that it can be used as the base atom-core element, as one now is able to explain with this element the internal, up to almost ten-dimensional (spacetime) atom-core super symmetries, and view with these base-core element symmetries the external electron-layered symmetries of all the elements listed in the periodic system, all isotopes included. The inverse hydrogen isotope is stable as a one proton element, having one extra, a third electron, but is unstable in symmetry as it is missing its opposite (two down, one up quarks) unit to complete the lepton-meson-baryon symmetries, which cause the proton/neutron particles observable in all stable elements. The production of the element deuterium, using four electrons, cannot be done in the relatively-flat, low-dimensional curvature value proto-sun. A higher, fatter dimensionally curved second-generation sun with stronger internal gravitational curvature values, meaning a deeper cauldron, is required. Deuterium shows the relatively-small massive quarks as a mixture of baryons, three-quark systems (*baris* is Greek for "heavy"), and mesons, two-quark systems (*meso* is Greek for "medium"), creating one proton and one neutron. Forming this way the deuterium atom core, you see with a locked-in matching electron cloud gives answer to the mass, the energy, the shape, and the (particle/antiparticle) charges of the internal atom bonds and retaining on the atom's surface the negatively charged, positively curved, interlinked layered electron cloud as a shield that protects the atom cores. Deuterium has the perfect base symmetry to describe all the other relatively-stable atom units and the relatively-unstable isotopes in the universe, looking exactly the way we observe all elements. The model clearly shows that at every gluing process to the next atom in the periodic system, less high energy paper or glass lepton' parts

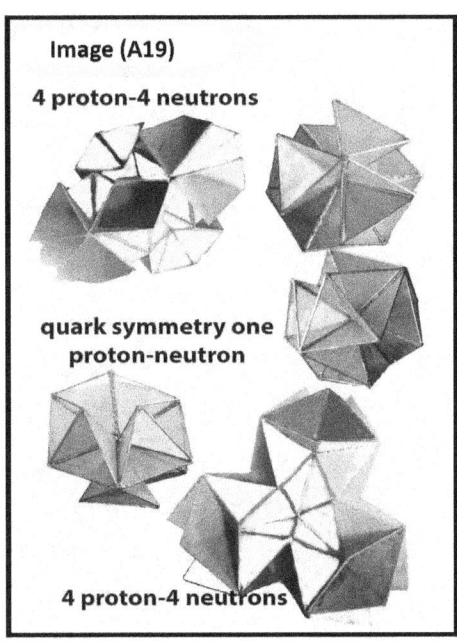

Image (A19)
4 proton-4 neutrons
quark symmetry one proton-neutron
4 proton-4 neutrons

are needed to build the next element symmetry model. See image A19. So it has at every step, surplus super-high energy spare to burn by giving off free high-energy electrons, and those, once released from the cauldrons inside, give off en masse on their outward journey, high energy as a stream of photons. Keeping the hard-working suns relatively stable and shining for a while also is showing the gravitational change from the first-generation, low curvature atom hydrogen, with relatively real, flat dimensional curvature values to the last stable generation atom curvature values of the element called lead, number 81, and then the relatively real, unstable fat dimensional curvature values that will produce all relative stable and all relative unstable radioactive isotopes up to the last isotope, element number 118. A relatively-low-dimensional first-generation proto-sun mostly produces the first inverse isotope hydrogen. A second-generation sun, a one solar mass sun, is needed to change the first-generation inverse hydrogen isotopes to deuterium and produce from then on, step by step, twenty-five elements up to iron (element number 26) and its isotopes. For larger elements, higher dimensional suns are needed. The higher the dimensional curvature value of a sun, the shorter is this sun's life span, and the deeper the sun's cauldron, the fewer yet heavier elements these suns are able to produce. Yet the different dimensional suns may possibly be converting the same many quarks in their life span. The first-generation proto-sun builds more atoms with less mass; the last-generation sun builds fewer atoms yet each with more mass, causing gradually denser elements with steeper internal dimensional curvature values up to the point where it is impossible to form a relatively stable or relatively unstable triangle of a no matter / dark matter bond. As you can see, once the hypotenuse of the dimensional (spacetime) object reaches the ninety-degree angle mark, no matter units and dark matter units both become inversely directed singulars. The no matter units go up and outward and the dark matter units go down and inward, so both cannot form interaction of any kind and so cleanly separate. Yet as the now mindless no matter unit moves high up from deep down, so does the mindless dark matter unit go deep down from high up. This is causing conception still deep down of a relatively real object, containing now a reciprocal computing device that in principle is capable of potentially thinking when forced in the need to survive. This you can explain with the internal resulting quark symmetry and super symmetries of the atom cores and also with the resulting external electron cloud symmetries shielding the atom cores, in two similar versions and with it form a logical reciprocal computing matrix similar to what the flat Fano plane matrix is. Yet unlike the flat Fano plane matrix symmetry, my basic quark atom-core model is showing four-sided and internally curving reciprocal, outside/inside (yes/

no) plus inside/outside (yes/no) computing grids in bond and so are way more complexly interconnected than the flatter Fano plane symmetries are. The Fano plane we know is able to compute, and so I think are the resulting internal and external connected symmetries of my atom-core model; therefore, these reciprocal symmetries are potentially also able to compute. We all have this potentiality as all atomic elements form in principle a reciprocal computer; yet innate objects cannot use this computing potentiality, having no sensors and having no sense naturally; they are not aware. We DNA-like objects in the universe do have sensors and so have the sense to use this computing potentiality; therefore, all DNA-like objects in the universe know to a degree how to think and how to make decisions as they gradually learn in the survival game to compute. They gradually change by doing anything that they can get away with, without getting hurt too much, and gradually evolving to produce relatively real, smart arses with relatively real big mouths.

Part IV

How Aware Are You?

Awareness, like everything else in the universe, is relative. Awareness is based on the ability to compare past observations to the now and that means to have memory. So I don't think that minerals, rocks, or mountains have the awareness of being a crystal, rock, or a mountain or have the ability to compare, meaning to observe, nor have any memory. The parts that make up the rock or the mountain certainly do have a so-called reciprocal computing potentiality as you have just seen with my atom-core model, yet I don't think that crystals, rocks, or the mountain have the needed sensors to have sense to know that they are a crystal, rock, or a mountain. I know that we all are in principle relatively similar, and I know that DNA causes our good looks and our potential health, and I know that it also causes our large powerful computer called the brain. Yet I also know that all events within the growing of that newly conceived, DNA-based baby may also have influence on the health and possibly has influence in the future mind-state of that baby. Yet I also know no matter what logical theory one may come up with, I can assure you that nothing ever exists twice. The mountains come and the mountains go as they gradually change from lava into rocks to gravel to sand to dust and to ashes. You were born when you were relatively young, and you will die when you are relatively old. The where, the when, or the why of your existence is pending on DNA, choices of drug use, eating, drinking, smoking habits, etc. or is related to the circumstances that happened to the mother and the father: war, famine, comet strike, flood, fire etc., which led to producing the new baby. Yet most of all, after the birth of the object, for example "you," it is the learning that will make you, "the you" who you could possibly be aware of, not the DNA that you have inherited. DNA describes only the biological part of you. DNA only builds the fancy hardware part of you that gives you the potentiality to be a relatively-real smart ass, if you want to be. We all are our own sculptors as we all do shape and change our bodies by exercising or by lazing about or by taking steroids or by lying on the couch and stuffing ourselves with fast food or by starving ourselves. All methods we use shape our bodies, and in the same way, we also shape our minds by exercising. For example when we survive difficult experiences out in the wild, in the office, or in some metal-working workshops or when we acquire knowledge by reading mountains of books or lose potential knowledge without thinking or while taking mountains of mind-blowing drugs or by willingly dumbing ourselves by lying on the couch and stuffing permanently our minds while watching TV, filling our minds with screen food. As I said, all methods that we use shape our bodies and so do all these methods also shape our minds. Awareness is the mental part of the ego, and we have to work hard to get that thinking part in our life's journey as no one can ever do the learning needed to make the potential mental ego that we mentally could be aware of if one would give it a go. Our awareness is relative and is of our own making, pending the inherited or otherwise acquired individual's hardware and/or the individual's willingness to think difficult or complicated thoughts. We also have the body potential to act on those thoughts. We only can inherit the relatively real computer part that we call the brain, and this brain is, in most objects, located in the fancy, dazzling-looking head that some of us use, if they work real hard to become real smart. Yet most of us only use this fancy, dazzling-looking computer to smile as a bright smile substitutes on TV, faith book, or in job interviews. If we are trained to do battle, of any kind, then we are more likely to be well aware of that kind of battle that we have trained so hard for so that we are more likely to survive this battle, be it sex, chess, war, or checkers. Since Darwin we know it is the survival of the fittest and so it was before Darwin. Some knew of it, and some acted on it; some others hoped for a merciful God and believed in orders, and some others went hiding. Yet in hindsight after the fight, we usually create excuses for our unfitness and inabilities to do battle and loudly claim compensation. Yet reality shows that any form of interaction is costly on both sides, so we always have to pay for wear and tear, one way or another when we interact and mostly accept the loss due to the pleasure of the resulting experience. Think of a supped-up racing car that wears its pistons out in no time while you are having fun racing or a house that constantly needs upkeep or else it ends up a garbage dump or the never-ending love battle of the sexes, which also needs upkeep or it too wears out in no time. The same may be true in our own mind engine, which creates the memories we need to make us individually, relatively conscious as me, or you, so that we are able to make relative sense of our reality. The by-ourselves created mind engine too wears out, but the mind engine wears out inversely to the bought car engine, which quickly wears out its pistons when heavily used and so the pistons have to be replaced regularly with new ones to keep the engine running. The brain engine is made of matter objects, yet the memories are not like the car pistons are. The homemade mind pistons are not made of matter objects and so the memories wear out inversely to the matter objects. That means that they gradually wear out when they are not regularly used as using and creating memories is as refreshing for the mind engine as fabricating and replacing the pistons is refreshing to the car engine. A well-maintained racing car engine enables great fun while racing that car and the same goes for the mind engine. We have all heard about Alzheimer's disease, and we know that it causes memory loss. We all have theories, but no one is certain as to what exactly causes this disease, yet we know it does alter or wipe out, some or all of our own, with acquired memories, and this shows that our memories may be similar to the pistons of the car engine and so may not be stable forever as absolutism is insisting our minds/souls to be. Many of us so deeply hope for it to be true. Memories very likely deteriorate due to not being refreshed or maintained as a relatively-real hologram. Our minds use something similar to a hologram or similar to the 3-D images we had about thirty-five years ago, where one had to hold a certain way to see the 3-D image or else it was just repetitive shapes. This image is produced on our so-called slate and is possibly equivalent to an electrically charged holographic-like image that regularly needs electrical charge and also needs to be kept shiny, like your cars paint job or else the surface of the painting on that slate will gradually fade in the same way as the paint job will fade on your car once you stop the polishing it. And so memories fade. The memories contained on the slate could also be damaged in accidents, just like your car engine could be damaged in an accident, or by some disease via prolonged alcohol and drug abuse akin to the use of the wrong fuel in the car engine or in any other combination that may cause memory loss, like power loss in the car. In the last twenty-nine years, at least with the introduction of Windows 95, you all have had some computing experience and you all know that computers crash and that data is lost—so you back up and back up the backup to protect your computer's memories. Unfortunately you cannot do the same to your mind. You cannot store your mind data somewhere as implied within a virtual cloud that sits, as some may think somewhere high up in heaven, where the data is meant to be safely protected by the angels. In reality, this so-called virtual cloud is a folder within one relatively-large server, meaning another computer. We all have lost audio and film data over the years as we have changed from film to video or from reel to reel to tape cassettes to digital disc or to solid-state memory drives. Data is lost. The records got scratched, the discs got warped or otherwise malfunctioned, and so info and memory was lost. With our own private memories, we know the rule of the thumb, use them or lose them. Like DVDs, memories are fragile as you the librarian, who always archives, evaluates and reads your memories may have used some obsolete memory space to overwrite with newer memory, and so older memory is fading. The symmetry that is describing the calculator called the

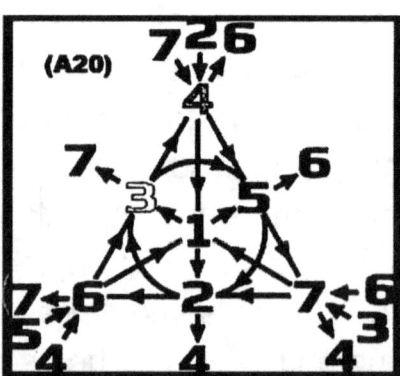

The Fano-Plane was invented about one-hundred and fifty years ago in Italy by a guy called Fano. This plane is capable to do a huge amount of computing using those symmetries.

Fano plane or Fano calculator—see image A20—and the symmetry we use to describe the light leptons, the resulting medium mesons, and the heavy baryons as two-quark and three-quark systems thereby forming the protons and the neutrons are so like the Fano plane symmetries on my atom model so that the displayed symmetries on my model are also able to compute yet in a more complex way. The relatively real dimensional (spacetime) symmetries on my barium atom-core model—see image A21—show five times four reciprocal, two-way, outside/inside plus inside/outside Fano calculator planes as a computing network, and you can see that the fifth part of this unit is the internal one. This internal one is shared by the four joining outer ones, ergo it becomes a virtual inner one, forming the core element. This now virtual inner core element may also be used in an computational sense as a

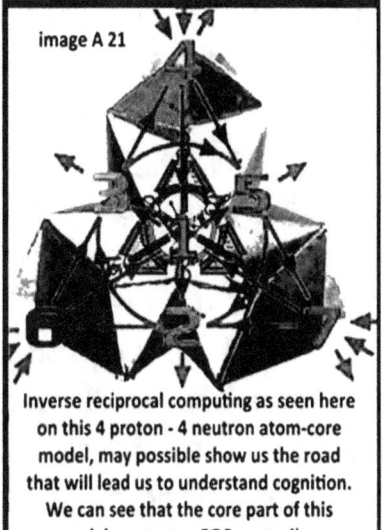

Inverse reciprocal computing as seen here on this 4 proton - 4 neutron atom-core model, may possible show us the road that will lead us to understand cognition. We can see that the core part of this model may act as EGO controller.

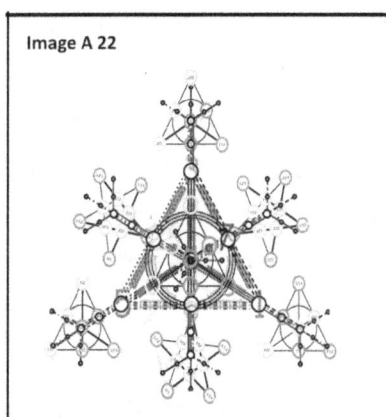

The Fano Plane seen here, is similar to the baryon and my meson atom model schematic that yet is not flat but form a sphere like object that gradually rolles-up with equivalent reciprocal inside-outside Fano Plane-like symmetries. These symmetries so I think may us allow to describe cognition.

relatively real thinking one, at least in theory, as this inner virtual core element is part of all the joined outer others and stays in contact in that position all the way from deuterium to lead as core controller by getting more and more complex with each transformation. Those relatively-flat atom-core elements already display a relatively high degree of complexly layered networking duality symmetries that have way more computing power than the flat, simple Fano plane calculator has. See image A22. In heavier elemental symmetries, the layered and networked reciprocal Fano calculator planes have a lot more complexity and have a lot more processing power. Those elemental layered networks are always interconnected in a fourfold reciprocal symmetry that gradually sprouts and twists in one direction, always having two matching and relatively real computing sides, except in the relative unsymmetrical isotopes. Yet only those objects that are able to operate their potential reciprocal computing ability are living things, made with DNA or DNA-like computing language, and only those can potentially be relatively aware. We all know that we always have to use hard work and make mistakes to get any form of noteworthy understanding, or awareness, of what one wants

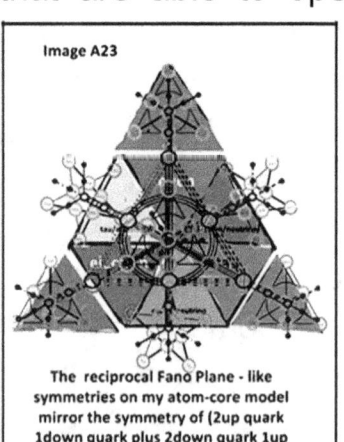

Image A23

The reciprocal Fano Plane - like symmetries on my atom-core model mirror the symmetry of (2up quark 1down quark plus 2down quark 1up quark) with-it form so I think cognitive computing.

to think and do. This is called learning. Understanding is hard work as it requires thinking, and all innate objects in the universe I have found do not learn or think at all, and of those objects in the universe that potentially can learn, most of them avoid it like the plague and use thinking only as a desperate last resort to survive and so will never, ever reach their mental potentiality to understand reality and how to do any good for themselves—let alone do any good for others. So without understanding, they fall into depression and so may commit suicide, or resort to stealing, robbing, and murdering others. Many of those who lack understanding and not resort to the above may consume excessively alcohol and drugs to cover up their lack of understanding and so live in the permanent belief of a magical fairyland in the hope that flicking the "Bier wand" (bottle) will make one understand.

The Fano calculator or Fano plane symmetries on my atom models allow me to think that the atom super symmetries are also able to calculate, in a similar fashion, like the Fano Plane does. See Image A23. The atom symmetries are layered from the low curvature atom values to the high curvature element values. This is shown in the element's external electron-layered shells and also in the matching internal quark super symmetries that form the protons and the neutrons. This reciprocal matter symmetry network may also be used as a relatively real programming matrix. This reciprocal, outside/inside (yes/no) plus inside/outside (yes/no), networked computer-processing matrix made of matter symmetry appears to have the logical and practical calculating and feedback ability to compare the reciprocal (yes/no) decisions. Those only indirectly observable atom symmetries form a coherently networked, outside/inside plus inside/outside, computing-software programming matrix. Both the surface Fano plane network in conjunction with the internal Fano plane network may well be all that is needed for DNA-based objects to have the reciprocal equivalence to operate a relatively real, relatively-cognitive computing-software programming duality. This symmetry network, being in reciprocal computing duality, is potentially also able to compare. To compare means to observe, to observe means to be aware and to be aware means to have memory. We all use temporarily in our lifecycle a relatively observable, relatively cognitive software-programming network when we think and do. We all do have the knowledge that we perform, with that by ourselves for ourselves created software program on our outside and on our inside only relatively well, pending our willingness, daringness, need, or abilities to learn and pending our sensors. Our own computer software programming intentions we process in both ways: with our inside relatively fast. We execute our computer-software programming decisions with our outside relatively careful and relatively slow, just to make sure we don't kill ourselves. With the reciprocal outside/inside plus inside/outside calculator software programming network, the temporal network controller, ego, is able to compare the old holographic-like recorded memories of our past experiences, which are stored on our inside, to the new experiences, which we are confronted with in the now, on our outside. This allows our temporal, relatively real calculator programming code controller, our ego, to understand by observing, thereby learn and judge with reason relatively easily. I am sure that cognition is also relative. Therefore cognition is only relatively achievable and is only relatively observable in all DNA-based objects. The emergent cognitive outcome on any relatively real object is pending on the object's created, complexly layered computing programming symmetry, its physical sensory ability, and the relative necessity to survive by writing a relatively-smart, individual, awareness computer-software programming code. This computing-programming unit, written with our own reciprocal programming language, appears to allow for relatively-real cognition in many, many different shades. Flora and fauna are living things yet do not use mobile phones or watch *Neighbours* on TV. I am sure plants or viruses have a much different level of awareness then you and I have, yet plants do have measurable awareness, different from anything else that we humans call living awareness. We also know that plants and viruses can evolve. Evolving means to change, and changing to suit one's survival needs means to be aware, each to their own. No one is the same; we all are individuals, and we all fight egoistically for our own survival, pending on our needs, as we all do what can be done to have as much pleasure, fun, or money, etc., as we can get away with in the relatively real survival game. Until we are stopped by unforeseen boundaries and then we whine, complain, and ask for compensation. As said, each to its own, one with more understanding and less greed, one with less awareness and more greed, yet all relatively-real, DNA-based living objects in the universe compute to compete in the survival game and therefore are to a degree relatively aware, pending on their needs and their abilities to survive. Yet when some wizards call the earth or a mountain a living, thinking object, then I am sure that the need and the ability to compute to survive for the earth or the mountain is quite different to the tree's needs or the virus's needs, yet the virus's needs and its ability to survive is not so different to our own need and ability to survive. The level of consciousness in all relatively-real living dimensional (spacetime) objects is pending, I guess, on each owns individual ability, willingness, daringness, and need for survival. We all do whatever we get away with as we explore our boundaries. And it is always the actual, yet virtual, programming code controller on the inside central core of the computing programming network, the ego, that records, evaluates, and routes our old outside/

inside plus inside/outside, computer-software programming code decisions on our networked slate. And those with (yes/no) evaluated, trusted old decisions try to influence us in our new intentions. I know that every experience shapes all of us. As well I also understand that these experiences are the tools for relatively real learning. I also know that it is the individual egocentric programmer who writes the sensory experiences relatively one way or relatively another onto one's symmetry-based, networked calculator, software-programming slate, thereby causing one's very own relatively real individual understanding and one's very own resulting memories, thoughts, desires, and fears. The algorithm produces one's own mind computer-software programming awareness we all have written with our very own private programming language symmetry. This private code is the result of our very own egoistic two-way (yes/no) evaluations of the sensory experiences that we had used to learn. Therefore no one will ever be able to read our mind symmetries directly, as no one will ever understand our very own individual programming language symmetries or our temporal egoistic code decisions, which are painted on our slate directly. You do not understand my temporal programming language symmetry painting or my codes directly, and I do not understand your temporal programming language symmetry painting or your codes directly. These individually changing software-programming language symmetries and the code meanings become vague due to being overwritten, or over painted, with new incoming sensory learning experiences, resulting in those earlier faulty symmetry code paintings to become forgotten or vague and so are mostly unconscious to the recorder, similar to giving a car a new paint job in a different colour. We never remember exactly the (yes/no) symmetry code painting we had used when we started to learn how to ride a bike or to swim for example. As we get better in riding a bike or with swimming, we overwrite the original painting and thereby hide the original faulty (yes/no) learning symmetries in the new painting. Communication and teaching between individual objects is therefore a difficult endeavour with the use of our common language. As for one, the first learning steps gradually become vague as we overwrite the older codes with our new learning experiences, and second each individual object makes individually their own different sensory experiences as we grow older and so attach our own private values as (yes/no) decisions to any experience we had encountered, pending on previous experiences. One handles a poisonous snake and gets bitten. Some other does the same and does not get bitten. Both do relatively the same, yet both have relatively different experiences and so create their own relative programming-language values and with them shape their own individualistic point of view of poisonous snake. We all have to learn in our own ways as no one else can do the learning for us. Teachers can only guide us along the way. Teaching and learning from others, in common language, is—pending all the involved individuals with their own acquired, individually learned experiences or circumstances and the ability to translate from the teachers individually learned programming language to commonly used language and then have the pupil translate into the pupil's own language causing the pupil's private understanding—relatively easy or relatively hard. All things are relative as all things constantly change, and therefore all events are individually and differently experienced and so are individually and differently evaluated. What is beautiful or good to me or you tonight may not be beautiful or good to me or you in the morrow. As we change so change our paintings, our memories, our desires, and our fears. Thereby we all are painting and over painting our own individual symmetry paintings for every experience we had evaluated and judged. Only learning cause's complexity in our gradually increasing awareness painting, therefore no painting of the one who learns will ever be the same, as the painting evolves, it is causing complexity. Complexity causes a greater understanding, however. Those who resist learning, or are in no need to learn, stick with the simple original painting they had acquired, in one way or another, and so are an artist who may not paint at all, or paint only a less complex pictures. One "want to be painter" paints with a few bold brushstrokes his memories and one master painter paints with millions of fine marks his memory paintings—each to its own. Those holographic-like symmetry picture paintings you paint form the awareness code for each individually experienced event. These egoistic (yes/no) symmetry paintings are also egoistic memory as those record the individually experienced events. Translating those different and individually changing language symmetry paintings and code meanings between individuals is causing fuzzy understanding. My most perfectly experienced harmonic symmetry paintings may cause the worst disaccorded chaotic experience to you. Relatively real awareness is potentially powered by the incoming singular dark matter. This one-way direction of incoming dark matter (always only from the outside to the inside) causes the one-way, dimensional (spacetime) energy flow in the actively-alive, relatively real dimensional (spacetime) object known as "us" and with it potentially allows the temporal and virtual computer-software programming code controller, ego, to compare on that one-way street, the past to the present and the present to the past. With it we are able to observe the changing flow of our once relatively-flat, relatively-mindless body to a massive old age, relatively-mindful dimensional (spacetime) body containing a lot of acquired, yet old, used-up energy and memories, if you are willing to observe yourself in the mirrors of reality. Comparing is observation, and observation enables relatively real understanding. The reason why I am not able to tell next week's lotto numbers is that next week's dimensional (space-time) object, the lotto numbers, have not yet received that in the future incoming singular dark matter part yet and so do not exist yet as a dimensional (spacetime) object. Therefore no one is able to observe the future, relatively real dimensional (spacetime) lotto numbers. So we all who like to play lotto must guess the future numbers. The accumulation of dark matter causes age and mass and with it allows for the massive memory potential from the past to the present in the sensory object, for example, you. Your physical ability and your mental willingness allow recording the new and allow simultaneously retrieving the old, in symmetry with your relatively real, networked, computer-software programming code controller: ego. We all, neutrinos, atoms, suns, planets, you and I, in the universe have been born with a relatively-empty programming slate. (There are many out there who think that the slate is clean at birth, and there are also many out there who think that the slate was filled by the will of God; both are wrong as both still think in the confusing infinitely-flat dimensional way. May I remind at this point again that reality is relative?) The relative little bit of the reciprocal (yes/no) programming code that is written on the tabular when we objects gradually start to be temporarily relatively aware and wake up at birth is likely caused by a relative, unconsciously acquired DNA memory painting. This memory painting also is overlaid by the relatively unconsciously acquired gravitational curvature image memory painting that the planets and the stars have painted onto the object's slate and is generally a snapshot recording at birth, causing a relatively-harmonic engraved picture duality on that slate with now external and internal symmetries that can be used as the core code programming guide to develop an individual's clever future programming plan and so paint this way a multitude of different objects, yet we can see this will never, ever produce the same painting again. The slate also may have been influenced with experiences, physically and mentally, from conception to birth. Most objects, like the neutrinos, electrons, atoms, molecules, planets, and suns, do not have the complexly layered sensory symmetries of the relatively-advanced, still relatively-unconscious, DNA at birth, therefore they do not think or understand at all. I know some wizards say that the electron knows when to give off and when to take on photons to balance their comfortable energy household, yet this is not so, as electrons take on and give off photons pending the room they reside in, not via knowledge. Those only have without the DNA sensors to observe the relatively-simple, relatively-unconscious recorded gravitational signature symmetry painting of the astrological influence as a dormant painting, not having an observing mind yet do have an engraved horoscope that is hinting on their future astrological potentiality, yet I am sure that electrons, atoms, houses, cars, ships, planes, or bikes themselves are not having the conscious experience to view or add to the basic slate painting nor observe the painting. None have any memory of it, not having DNA-like based sensors. All relatively-real, relatively (low to high) dimensional (spacetime) objects, like dark suns, neutrinos, electrons, atoms, molecules, cells, and bodies like you and me, have an outside/inside abacus of very different complexity. Those individuals all have potentially a computer-software programming capability and may use it if they can, need, or are willing to do so. Computers are not aware that they are computing; only the operator/programmer will learn to paint a symmetrical pattern that allows recalling those learned events. And so only this individual operator, the individual egoistic DNA-based code writer / controller, will know the meaning of the symmetrical-patterned painting—no one else. When the temporal, relatively real, relatively-conscious networked software-programming code controller dies, the programming-code controller and the memories crash, recording and a simultaneous observation of the recorded paintings ends. There cannot be relatively real after death. Cognition of the old recorded paintings as the incoming singular dark matter, which provided the dimensional (spacetime) energy flow to sustain the relatively real dimensional holographic-like memory paintings, has halted that energy flow and the relatively real holographic-like symphony painting is no more. The only temporarily, relatively real, relatively-conscious, computing-software programming code controller, the virtual ego, which controlled your relatively-alive awareness codes and views, the "by us for us" recorded paintings, has lost the incoming dark matter energy flow needed to sustain the multilayered reciprocal and relatively real symmetry painting and is thereby not only cut off from memory and cognition, it is being without awareness or memory and so is no more alive or conscious. Therefore we are not anymore relatively real and so are not anymore able to compare the past with the present or the present with the past. Death has stopped this mindful dimensional (spacetime) thinking. Death is the ending of our active relatively real, dimensional computing and programming life. The remaining molecules, atoms, and electrons are now dispersing, living their own individual relatively real duality without a mind.

Part V

Conclusion

I hope that you now are able to understand reality with observable, understandable common sense and simple Muggle thinking. My tale has shown that reality can very well be explained without the need to twist our observations with magical fripperies or the use of imaginary, infinitely-flat dimensional mind-farts—and likewise without the use of space-time matter singularities or with Theo-illogical and quite hot big bangs involving holy virgin angels that create life nor with mind-farted cold big crunches, involving evil horny devils that end life. I have not violated the observable laws of physics with the observation that I made in Oz-land: that neither space nor time can be dimensions but result of none dimensional singulars—one of no matter and one of dark matter, which in bond, cause the yin-yang, relatively real object now as a low-to-high dimensional (spacetime) object duality and is now relatively measurable with dual positive/negative gravitational object curvature values and positive/negative energy values. We all know that infinities have no dimensional values and therefore cannot be measured, and I am sure that all relatively thinking minds will agree with that, maybe not at first as some I have noticed seem to be quite stubborn, yet they will come around that the dimensions here on earth too, same as in Oz, belong to any relatively observable object, not to impossible, imaginary, confusing, infinitely-flat dimensional fairy tales. Those are enshrined with Theo to dress up smelly illogical mind farts. The use of the inversely directed no dimensional singulars of no matter going from the inside to the outside and dark matter going from the outside to the inside are mirrored with the positive/negative gravitational curvature values and are also are mirrored in the positive/negative energy levels and explain the dualities of particle/antiparticle symmetries that we observe as bonded leptons, causing quark super symmetries and with that produce one proton and one neutron as two up, one down quark plus two down, one up quark in bond as low-to-high dimensional (spacetime) objects solved all of the problems that you wizards are still encountering with the magical and so very hot big bang. We in Oz do not need the magic that the currently used model here on earth needs to trigger the nil point yet so very hot big bang event, which is describing with Theo-illogical magic the matter forming process nor do we use in Oz angel-like or virtually massed particles, which you still use in the hope of explaining the evaporation of a point-like space-time matter singularity. In Oz the relatively real dimensional (spacetime) matter-separation process is not evaporating matter via half massed or virtually massed space-time matter angels; in Oz the (high) dimensional (spacetime) matter object is separating its duality status into inversely directed singulars. We see that the no matter unit value is gradually going up, and we see that the dark matter unit value is gradually going down. Euclid has to be forgiven for not knowing that the earth is a gravitational sphere and so was excusable yet wrongfully thinking when he invented the three infinitely-flat space dimensions. Since Albert Einstein you boldly, yet wrongfully, think that you have the three theoretical infinitely-flat space dimensions next to the theoretically also infinitely-flat time dimension. You also think that you have the four singular dimensions in bond as a three-to-one or four-dimensional space-time object, yet Einstein should have known that all of reality dimensions included need to be relative to explain relativity, meaning that Einstein and his first wife, Mileva Maritsch, should have thought that the dimensions belong to objects, not to invisible infinities. Objects are formed with the no matter / dark matter bond, causing dimensional (spacetime) object dualities. Singulars are not objects and so have no dimensional values of kilometres or seconds to be measured with. In order to form a logical theory of relativity, space and time can never form singular dimensions, as dimensions belong to all relatively real dimensional objects, not to mind-blowing infinities. No dimensional no matter on the other hand and no dimensional dark matter you can easily see can gradually form objects once in duality bond and so explain the logical production of all the relatively real dimensional (spacetime) matter that you observe as the universal background radiation. You also can see that this background radiation is made mostly with relatively-flat and some relatively-fat dimensional (spacetime) spherical matter objects, all having dual gravitational curvature values between relatively low to high dimensional (spacetime) object curvature values. With both singulars now in bond, you have the ability to describe with equivalence and elegance in one story the observable inside and the observable outside of the universe. Einstein needed two versions of relativity and two versions of infinitely-flat dimensions to describe reality with and so had no option but to fail in the description of reality without using the logic embedded within equivalence to dimensions. Einstein used the three infinitely-flat space dimensions Euclid had invented and the one infinitely-flat time dimension that he himself had added as one individual singular type of dimension. Separately Einstein had used the four singular dimensions combined as a four-dimensional space-time, describing with those bonded four dimensions the resulting yet so confusing matter objects. Since Kaluza-Klein you have tried to imagine matter in five space-time dimensions. The modern version of the Kaluza-Klein theory tries to push imagination to eleven space-time dimensions. There is no way that anyone in the universe can imagine more than three theoretically infinitely-flat space dimensions in ninety degree angles to each other, and those theoretically infinitely-flat space dimensions you have seen are unable to describe the many dimensional curvature values needed to model the complex super symmetries that you observe in the high curvature value atoms in a logical way. Neither are those dimensions able to explain the relatively-large objects that we in Oz too observe in our common universe nor are these flat dimensions able to model the universe equivalently nor the observable inverse dimensional curvature values of our massive planets. Remember my thought experiment. Nor do those flat dimensions model energy. Calabi-Yau have attempted with the Calabi-Yau model—see image A24—using the thought concept of string theory to roll up many infinitely-flat dimensions to model the super symmetries of the atomic cores, first with five then six then eight and then eleven space-time dimensions. With those infinitely-flat, rolled-up dimensions, Calabi-Yau attempted to explain the needed high-dimensional curvature

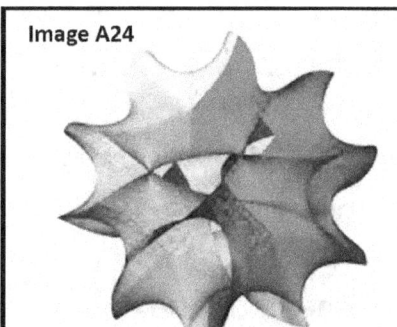

Image A24

The by Calabi-Yau theoretically rolled-up 'infinitely-flat' dimensions are very similar in shape and complexity to my own atom-core model yet my atom-core model has no need to roll-up 'infintely-flat' dimension as the, relatively real dimensional (spacetime) objects are not at all 'infinitely-flat'.

values observable in the atom cores. The space-time dimensions that you are still using to explain matter are impossible, are not logically thinkable, and are only usable to satisfy imagined flatland mathemagical theory and the wizard's egos. Those are leading to fairy tales, like to the fancy quite hot big bang nil point singularity or the quite cold big crunch nil point singularity. Also this flat-dimensional thinking is leading to impossible Hawking-Penrose point-like space-time singularities and weird wormhole-travelling fantasies, which open gateways into parallel universes that allow wizards like me to come to you with the flick of my wand. The so many independent dimensions that you also fancy in your thinking allowing fairy tales like travelling in hyperspace across the universe. It also allows teleportation fancies, like, "Beam me up, Scotty." As well it is leading to wacky time travelling mind farts, like travelling *Back to the Future* or travelling back to the past so you can kill some siblings of your great grandparents and this way vanquish in the present rivals and inherit their fortune. It's giving desire that you may get a lift in a Tardis with a time lord and travel into the future, check out next week's lotto numbers, and come back and play cheat lotto. You cannot travel in the singular kilometres of space, and you can never travel back or fast-forward in the singular seconds of time, as kilometres and seconds exist in the dimensional (spacetime) duality bond only. The singular space and the singular time do not exist, as they have neither the room nor the substance to grow or age, neither be measured nor judged. No matter and dark matter are invisible, are not-dimensional, are inversely directed mindless singulars. Singulars cannot compute as singulars are without form to construct a computing matrix, are not tangible and so have not the triangular computing room needed to make a yes/no decision, yet once in duality, the kilometres and the seconds needed for the relatively real dimensional (spacetime) computing grids to be a relatively real object. Neither the kilometres nor the seconds ever exist as singulars. Singular dark matter and the singular no matter on the other hand cause in bond the dimensional (spacetime) object curvature values needed to allow us to describe the object's mass and energy with (kilometres per seconds) and possible inversely also with (seconds per kilometres). We do need both no dimensional inversely directed singulars in bond to calculate and/or to measure the relatively-real dimensional (spacetime) objects, never just one. Imagined dimensions as infinitely flat or infinitely large are figments of the mind. No dimensional no matter is only singular, deep down below the relatively (high) dimensional (spacetime) container objects, where the triangle's hypotenuse, as it reaches the ninety-degree angle mark, is no more. The triangular object collapses, folding the energy duality potential to up and down and so is separating the (energy/mass) duality potential and becoming inversely directed mindless singulars. The singular, now free no matter is travelling upward and outward and is being penetrated on its outward, upward journey with the still incoming downward and inward flow of singular dark matter at the ninety-degree angle mark and so must have a head-on collision with incoming dark matter, causing conception, ergo a new relatively-real (low) dimensional (spacetime) object, the future neutrino baby object, describable now with an outside diameter potential of 599,999 kilometres yet is confined within a macaroni-like, funnelled container and travelling in one second upwardly, almost 299,999 kilometres. The moment that the neutrino baby was conceived deep down on the inside of the dark sun's container at the ninety-degree angle mark with a 0.00000001 dimensional (spacetime) curvature value, the neutrino will start to grow older in the gradually widening to relatively real, relatively (0.50) dimensional (spacetime) container object curvature values, with the still incoming dark matter. This accumulation of dark matter is potentially shrinking the neutrino's outside diameter by increasing internally its curvature value as mass/energy duality, gradually up to a $_{0.50}$ dimensional (spacetime) neutrino object, giving the potentially relatively-flat neutrino the outward push equivalent to its theoretical space value and so explains with logic the birth of the neutrino and shows that the neutrino's dimensional (spacetime) value is able to inflate the universe with a little bit faster than light, at least up to the ergo sphere, when this theoretical no matter value expansion speed is checked with the theoretical internal gravitational dark matter value, causing relatively real dimensional (spacetime). The relatively real neutrinos therefore are inflating the universe from the inside to the outside in all outward directions with objects, not with space and time dimensions, creating the universal background room that now causes observable radiation pending on density. No object like a so-called spaceship can travel in space, as space is not a dimensional thing and so does not exist, not in the relative, directly observable universe and not in the relative, indirectly observable parts of the universe. We can see with this underlined triangle that the relatively real gravitational hypotenuse only exists between the 0.001° angle and the 89.99° angle and so very, very deep down inside of reality at the ninety-degree angle mark is gradually becoming inverse directional. The now separated part of singular no matter is going up and the separated singular part of dark matter is going down; both are then inversely directed. That so-called spaceship is a relatively real object. This object is made with individual (2 to 6) dimensional (spacetime) matter object curvature values; it has overall relative gravitational curvature values. This ship travels in the observable, relatively real (0.50 to 1.40) dimensional (spacetime) background rooms, pending the room's curvature type and the travelling object's curvature value type—flat or fat—relatively easily or relatively hard. With this insight you can now clearly understand why relatively-flat (0.50) dimensional (spacetime) neutrinos are able to pass almost undetected through your relatively fatter (2 to 6) dimensional (spacetime) earthly matter. You can also understand why relatively high energy neutrinos interact a little more; you also can understand why larger objects, like photons or apples or lead balls interact, pending opposing medium quite a lot—for example when photons encounter air or water or high-speed depleted uranium darts encounter steel objects, brick walls, or you and me. We also can understand that when photons or apples collide with you or me or the earth's surface and exchange their curvature values, or parts of it, why you and I and the earth's surface are then heating up. This also explains why the relatively-fat (2 to 6) dimensional (spacetime) vessels, as NASA had observed, are losing constant velocity travelling speed in the relatively-flat regions, which may only have 0.50 to 1.40 dimensional (spacetime) curvature values, near and outside of your relative (2 to 3) dimensional (spacetime) curvature values, as those vessels you can now see, do not travel in dimensions of inherently-flat space, as you had thought they do, where the Voyager vessels may have encountered a unknown force, yet they do travel in relatively real (0.50 to 2) dimensional (spacetime) object curvature values, where they encounter, with their relatively-fat (3 to 6) dimensional (spacetime) object curvature values. Those smaller dimensional curvature values cause friction to the fatter object dimensional curvature values. Remember in Euclid's geometry a 3-D object does not fit into a 2-D surface or into a 1-D line, and likewise in Oz-land's relatively-real geometry a 3 to 6 dimensional (spacetime) object curvature value cannot travel in the relative 0.50 to 2 dimensional (spacetime) object curvature value with relative ease; yet relative 0.50 to 2.00 dimensional (spacetime) curvature value objects can travel relatively easily in relative 3 to 5 dimensional (spacetime) object curvature values. Therefore those vessels now are slowing down in constant velocity travelling speed as the relative three-to-five-dimensional spaceship is now trying to squeeze through those flatter 0.50 to 2.00 dimensional (spacetime) rooms with a much fatter belly—the same constant velocity travelling problem the aliens have. This also explains why you have never met any aliens as the energy consumption that the ship's motor would need to sustain the needed traveling speed of a relatively real fat dimensional (spacetime) ship that is permanently trying to squeeze through the relatively-flat dimensional (spacetime) rooms within any galaxy is impossible to sustain for such a long journey. The regions that lie between the galaxies are having far less dimensional curvature values and are much flatter and so travelling in those regions is, energy-wise, a mind fart, unless you are a no-rest mass particle, like a photon or a mildly massed neutrino. Besides the next door neighbour planet in our galaxy that is similar to your own planet and so also have alien wizards like me and with an environment that is suitable to your own liking is astronomically, one thousand four-hundred light years away from earth. Without the use of magic wands, thereby it is unthinkably far away to travel for a relative four-to-five-dimensional curvature value object. You also can now understand why the relatively lower dimensionally curved objects are accelerating on their journey into relatively higher dimensionally curved objects—think of Newton's apples—pending gravitational attraction (dimensional curvature value), relatively slow, or relatively fast, with one point six three meters per second per second on the surface of your moon, nine point eight one meters per second per second on the surface of your earth, thirty thousand kilometres per second per second on the surface of a one solar mass dwarf, and with two hundred ninety-nine thousand nine-hundred and seventy-two kilometres per second per second on the surface that I call the event horizon of the relatively real (9.50 to 9.99) dimensional (spacetime) container object called the dark sun. With those relatively real dimensions, you can visualize the very, very; very slow lifecycle of the objects in the universe, without being confused. You can understand the forming of temporal, relatively real dimensional (spacetime) containers with the head-on collision of no matter with dark matter, causing the duality bond and thereby gradually forming a relatively real, 0.01 to 0.50 dimensional (spacetime) object bond that causes the relatively-flat-dimensional objects that are expanding in all outward directions and give an explanation to the background radiation that Penzias and Wilson in 1929 had observed. Now you have a logical explanation to the relatively-large, observable expanding cosmology that Edwin Hubble in 1924 had observed, and with it you now can explain away easily the hot big bang and the cold big crunch mind farts. You can visualize the very, very small in high curvature internal quark super symmetries of the atom cores to the still relatively real yet relatively-fat (9.99) dimensional (spacetime) object curvature values of a dark sun. The external layered electron shells and the internal quark super symmetry structure of the element cores can provide answers to powerful reciprocal outside/inside (yes/no) plus inside/outside (yes/no) computing. With the networked, relatively observable code controller, ego, which controls outside/inside (yes/no) plus inside/outside (yes/no) computing, one can understand the resulting relatively real observable software programming code controllers as objects with relatively-smart cognition in many different shades. Just look around and you will see. This story explains the perpetual one-way flow of the changing relatively real, inversely directed singulars between 0.01 to 9.99 dimensional (spacetime) object curvature values. The theoretical null dimensional state in nought-degree angle and the theoretical ten-dimensional state in ninety-degree angle are right-angled to each other and so can never form a head-on collision to conceive in this state from the outside to the inside. Both singulars in that state are formless and thereby unable in this state to form a triangle. Therefore they are without the ability to compute nor have a mind, as the potentiality to compute is only contained within the duality of the dimensional (spacetime) hypotenuses triangle, and this triangle can only form via the head-on collision. This event happens deep down in the dark sun's container, at the ninety-degree angle mark, where a duality via a head-on collision from the inside to the outside motion of no matter and the inverse motion of dark matter is causing conception, and so the dimensional (spacetime) room that allows the penetration of dark matter from the outside to the inside in all relative observable objects of reality causing mass and age within any dimensional (spacetime) object and with it change their dimensional values. You can see that once no matter and dark matter are inversely directional, singular no matter goes up, causing on its way the head-on collision with dark matter, which is on its way down and so causes the conception of a new (low) dimensional (spacetime) object with up-quark and down-quark dualities forming atoms. Dark matter is now able to go down into the object from the outside to the inside, causing age and mass and dark suns. No magic needed.

And now it is goodbye as I now have to go. I, Adremolin, the last visiting Wizard of Oz, is now trying to use his magical wand one more time to catch the last magical wormhole opening before it closes up. As you see magic is fading too here on earth, and I really want to catch that last ride back to my Oz-land home to forge another story.

Bibliography

The Future Eaters Tim Flannery

The Bible Code Michael Drosnin

Quarks urstoff of unserer welt Harald Fritch

Frontiers of Complexity Conveney/Highfield

The Descent of Mind Evans/Deehan

The Chalice and the Blade Riane Eisler

Guns, Germs, and Steel Jaret Diamond

The Celestine Prophecy James Redfield

The Elegant Universe Brian Greene

Fuzzy Thinking Bart Kosko

The Miss Measure of Man Stephen Jay Gould

Bully for Brontosaurus Stephen Jay Gould

The Sacred Balance David Suzuki

The Theatre or Mind Henryk Skolimowski

The Large, the Small, and the Human Mind Roger Penrose

The Road to Reality Roger Penrose

The Emperor's New Mind Roger Penrose

Shadows of the Mind Roger Penrose

Chaos James Gleick

The God Effect Brian Clegg

Mathematical Mystery Tour A. K. Dewdney

How the Mind Works Steven Pinker

God's Secret Formula Peter Plichta

The Holographic Universe Michael Talbot

The Meaning of All Richard P. Feynman

Black Holes and Time Warps Kip S. Thorne

Before the Beginning Martin Rees

The Mind of God Paul Davies

The Last Three Minutes Paul Davies

Super Force Paul Davies

Space-Time and Beyond Fred Alan Wolf

Parallel Universes Fred Alan Wolf

Neuland des denkens Frederic Vester

Introduction to Symbolic Logic B. O'Conners

Climbing Mount Improbable Richard Dawkins

The Selfish Gene Richard Dawkins

River out of Eden Richard Dawkins

The God Delusion Richard Dawkins

The Extended Phenotype Richard Dawkins

The Greatest Show on Earth Richard Dawkins

The Blind Watchmaker Richard Dawkins

The Blank Slate Steven Pinker

Einstein Philipp Frank Stephen Hawking

The Universe in a Nutshell Stephen Hawking

A Brief History of Time Stephen Hawking

Black Holes and Baby Universes S. W. Hawking

Stephen Hawking: A Life in Science John Gribbin

In Search of the Double Helix John Gribbin

Newton William Rankin

The Origin of Species Charles Darwin

The Structure of the Universe Paul Halpern

Fermat's Last Theorem Simon Sing

A New Science of Life Rupert Sheldrake

The Present of the Past Rupert Sheldrake

Infinity and the Mind Rudy Rucker

At Home in the Universe Stuart Kauffman

The Cosmic Blueprint Paul Davies

Ripples on a Cosmic Sea Blair/McNamara

The Tao of Physics Fritjof Capra

The Wheel of Life Fritjof Capra

Five More Golden Rules John L. Casti

Mathematics: The New Golden Age Keith Devlin

The Fabric of Reality David Deutsch

The Mind of the Machine Kevin Warwick

The Collapse of Chaos Terry Pratchett

The Origin Irving Stone

www.ingramcontent.com/pod-product-compliance
Lightning Source LLC
Chambersburg PA
CBHW071417210526
45465CB00001B/426